Lighting for Health and Safety

T0259021

Lighting for Health and Safety

N.A. Smith BA, PhD

Honorary Research Fellow,
Institute of Occupational Health,
University of Birmingham, UK

Routledge
Taylor & Francis Group

LONDON AND NEW YORK

First published by Butterworth-Heinemann

First published 2000

This edition published 2011 by Routledge
2 Park Square, Milton Park, Abingdon, Oxon OX14 4RN
711 Third Avenue, New York, NY 10017, USA

Routledge is an imprint of the Taylor & Francis Group, an informa business

British Library Cataloguing in Publication Data
A catalogue record for this book is available from the British Library

Library of Congress Cataloguing in Publication Data
A catalogue record for this book is available from the Library of Congress

Composition by Scribe Design, Gillingham, Kent

ISBN - 978 0 7506 4566 9

Contents

Preface

The late 1800s saw gas lighting installed in domestic premises and at the beginning of the twentieth century, with the advent of electric lighting, homes and industry were able to continue with visual tasks during the hours of darkness which hitherto had only been possible during daylight.

Artificial lighting has become so commonplace that it can sometimes be taken for granted and therein lies a possible minefield of related health and safety problems.

This book aims to guide the reader through the fundamentals of vision and lighting, to highlight the potential health and safety problems that can develop as a consequence of inadequate lighting and further to advise of the necessary remedies available in order to produce optimum lighting conditions for the workplace and therefore assist the practitioner in compliance with legislation.

N.A. Smith
August 1999

Acknowledgements

Relatives and colleagues have provided assistance in the preparation of this book; to each and every one of them I extend my sincere gratitude.

It would, however, be remiss of me not to single out the contribution of one of my colleagues, Dr Sandip Doshi. His guidance and constructive advice with the preparation and development of Chapters 4 and 5 has been invaluable, and to him I offer special appreciation.

N.A. Smith
August 1999

Chapter 1

Introduction

1.1 History of light sources

From the earliest of times right up until the nineteenth century, the production of artificial light almost invariably involved the use of flames. Some of the major problems encountered with naked flame light sources were:

- the method of production of the flame;
- the means of keeping the flame burning for long periods without the need for constant attention;
- the attendant fire hazard problems.

Surprisingly, even in the middle of the twentieth century, a large proportion of the population of the world were still using flames as the principal light source in the home.

In order to produce light by heat, forms of combustion are necessary. Such methods of combustion used over the years have ranged from primitive forms of flame torches and candles (wax and tallow) to oil lamps and gas mantles. From the Middle Ages through to the nineteenth century, domestic lighting often incorporated *rushlights* These were made by stripping a rush stem of its very thin rind which was then repeatedly immersed into a hot fat until a suitable thickness was achieved. Up until about the sixteenth century, candles rarely found a use in domesticity, except the houses of the affluent, and were used predominantly for religious purposes. Candles still play an extremely important role in church tradition: February 2nd is Candlemas Day. This festival celebrates the presentation of the infant Jesus by Mary in the Temple at Jerusalem. There she met Simeon who recognized Jesus as a 'light to lighten the gentiles' (St Luke 2.32).

One early form of lighting frequently used for the lighting of theatres involved the use of burning quicklime until it reached incandescence and emitted visible radiation. This method of lighting led to the often used phrase 'in the limelight'. This form of light production was, however, not without its problems, including that of fire.

During the eighteenth and nineteenth centuries artificial lighting went through a period of tremendous excitement. The burning of gas obtained from heated coal paved the way for a new form of lighting and in 1792 William Murdoch installed gas piping and jets in his property in Cornwall. In 1827 the friction match was invented; prior to this time tinder boxes containing flints, steel and wood splinters tipped in sulphur were used in order to start a flame. In 1860 mineral oil was turned into paraffin which was used for lamps and for the manufacture of wax candles.

In 1885 Count Aver von Welsbach began investigating the characteristics exhibited by rare earths when they were heated. In 1893 he developed the incandescent mantle where the production of light is not dependent upon the luminosity of the flame produced but upon the level of incandescence to which the mantle is raised.

There were major problems associated with the use of gas for lighting. It was both poisonous and explosive and incidents involving leaks and other malfunctions often led to catastrophic consequences.

In an attempt to provide other means of artificial lighting, there had been some experimental work with the use of electricity. In 1810, and whilst gas lighting was still the dominant source, Sir Humphrey Davy had shown that an arc could be established following the separation of two touching carbon rods which formed part of an electrical circuit. The arc lamp was not established in widespread use, however, until the 1850s. The quantity of light produced by the arc coupled with the requirement for a suitable electrical supply precluded its use in domestic installations. It was used predominantly for public lighting, building lighting and in lighthouses. As with the earlier light sources which were totally dependent upon the production of flames, arc lighting was messy to maintain.

Michael Faraday's discovery of the principles of electromagnetism in 1831 paved the way for electric lighting in its present form. The first practical electric lamp was the incandescent filament lamp produced almost in parallel in 1879 by Swan in the United Kingdom and Edison in the United States. The first domestic electric lighting system using filament lamps was installed in 1880 by Lord Armstrong in Northumberland, the lamps being supplied by a water-driven dynamo.

The production of artificial light by electrical discharges was achieved almost by accident since the early scientists were primarily concerned with the manner in which electricity was conducted through a partial vacuum.

Peter Cooper-Hewitt introduced his mercury-arc lamp in 1901. The mercury was contained in a sealed discharge tube and in order to start the discharge, the lamp had to be manually tilted so that the mercury would fall down to the lower electrode and the arc would establish.

Early attempts at commercial discharge lighting met with limited success and it was not until the early 1930s that lamps were produced which found universal approval, these being based upon the use of sodium and mercury. By the late 1930s fluorescent lamps were being produced.

Improved technology has allowed major advances to be made in lamp production, for example high pressure sodium lamps with their greatly increased colour rendering properties have been developed. The metal halide lamp, a progression from the basic high pressure mercury vapour lamp, has also been introduced.

An innovation in lamp technology is the electrodeless or induction lamp which relies upon both magnetic principles and fluorescence for the generation of artificial light.

What of the future? Scientists may well consider that attempts to increase the luminous efficiency of those lamps presently available should be investigated. Alternatively, lamps dependent upon the use of elements other than mercury and sodium may be further developed and become commercially viable.

Chapter 2

Fundamentals

2.1 The electromagnetic spectrum

Light is a form of energy. It passes from one body to another and can do so without the need for any substance in the intervening space. Such energy is termed radiation and it is said to be electromagnetic in character. The radiation thus has both an electric field and a magnetic field. Both of these fields vary sinusoidally and are mutually at right angles, as shown in Figure 2.1.

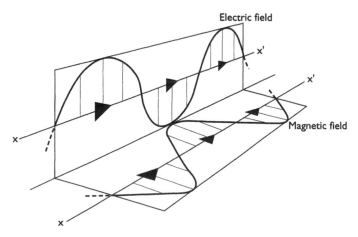

Figure 2.1 Electric field and magnetic field mutually at right angles. Electric field and magnetic field have same axis x–x' but are shown separately for clarity only.

Visible radiation is the term given to that radiation which is detected by the eye. It occupies only a relatively narrow range of wavelengths within the whole of the electromagnetic spectrum. Figure 2.2 shows the electromagnetic spectrum with the visible spectrum shown in detail.

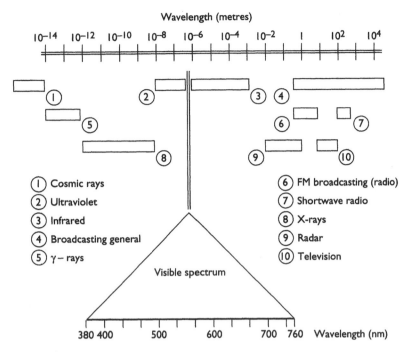

Figure 2.2 Electromagnetic spectrum.

2.2 Optical radiation

Electromagnetic radiation which lies in the wavelength range 100 nanometres to 1 millimetre is referred to as *optical radiation*. It follows that optical radiation covers visible radiation together with ultraviolet (UV) and infrared (IR) radiation. Whilst the visible spectrum is essential for seeing, UV and IR radiation can have adverse effects on the body, including the skin and the eyes.

All bodies emit optical radiation over a relatively wide wavelength range. For a black body (see Section 2.5), the peak emission of radiation occurs at a wavelength which is determined by Wien's law:

$$\lambda = \frac{2\,898}{T} \tag{2.1}$$

where λ is the wavelength (in micrometres) and T is the absolute temperature on the Kelvin scale.

2.3 Ultraviolet and infrared radiation

At lower wavelengths than the visible spectrum the radiation becomes ultraviolet (UV), whereas at higher wavelengths than the visible

Table 2.1 Classification of UV and IR radiation

Group	Ultraviolet	Infrared
A	315–400	780–1400
B	280–315	1400–3000
C	100–280	3000–1000000

All wavelengths are given in nanometres.

spectrum the radiation becomes infrared (IR). Both IR and UV radiation are sub-divided into three groups as shown in Table 2.1. Ultraviolet radiation is usually produced either by the heating of a material to incandescence or alternatively by the excitation of a gas discharge. The major source of UV radiation is the sun, which can be considered to be a huge incandescent mass. When produced by incandescence, UV radiation is in the form of a continuous spectrum.

An everyday example of the emission of ultraviolet radiation in industry is the electric arc produced in the welding process. The arc is established by the passage of an electric current across an air gap and between two metallic conductors or electrodes. The electrode tip, workpiece and airgap are collectively heated until incandescence is reached. Analysis of the spectrum so produced will show that there is a continuum with discrete spectral lines superimposed. The characteristics of these lines are influenced by the properties of the materials from which the electrode and workpiece are constructed and by the properties of the surrounding gases.

2.4 Continuous and discontinuous radiation

When a solid object is heated to a sufficiently high level it reaches incandescence and the electrons will be violently agitated resulting in constant collision with other electrons. Some of the energy which results as a consequence of these collisions will be radiated from the hot body. Because of the closely-packed arrangement of the individual atoms in a solid body, the energy released will appear as a continuum characteristic of the temperature of the radiating body. The output is therefore radiation which is smoothly distributed over a relatively wide range of wavelengths. Some of the output wavelengths will be in the visible range but by far the greater proportion will appear in the infrared region as shown in Figure 2.3. A gas or vapour, when excited, behaves differently to a solid body. When excited, a gas or vapour will contain electrons which have been raised to a higher energy orbit around the nucleus. These electrons will very quickly return to their normal orbits and in so doing the energy they release is emitted as photons. Such radiation is usually emitted at a sequence

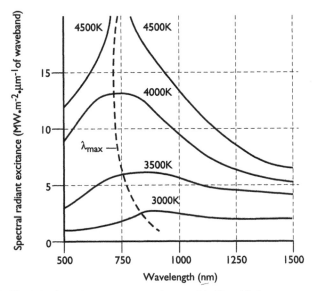

Figure 2.3 Black body radiation in accordance with Planck's law.

of one or more wavelengths producing a discontinuous spectral output.

2.4.1 Spectral power distribution (SPD)

This is the term applied to the plot of variation in output from a body, typically a light source, with wavelength. The curves of Figure 2.3 are spectral power distributions (SPDs).

2.5 Black body radiation

A black body is one that exhibits the maximum amount of radiation theoretically, and completely absorbs all incident radiation which falls on it when the body is maintained at a uniform temperature. There are no such bodies in reality although many bodies closely approximate to it. If any body is held at a uniform temperature it will radiate thermal radiation in accordance with Planck's law.

Figure 2.3 shows the relative energy radiated at various wavelengths by a black body at different temperatures. This shows that the wavelength of maximum radiation decreases as the temperature increases. In addition, Wien showed that the product of wavelength of maximum radiation and absolute temperature is a constant.

2.6 Wavelength, frequency and the velocity of propagation of light

Light travels sinusoidally in waves as discussed in Section 2.1. A relationship exists between the length of the wave, its frequency and the velocity of propagation such that:

$$\begin{array}{c} \text{Velocity} \\ \text{(metres per second)} \end{array} = \begin{array}{c} \text{Frequency} \\ \text{(Hertz)} \end{array} \times \begin{array}{c} \text{Wavelength} \\ \text{(metres)} \end{array} \qquad (2.2)$$

2.7 Radiant flux and radiant efficiency

Radiant flux is the term given to the total power of electromagnetic radiation (measured in watts) emitted or received. The value may include both visible and invisible components, the visible component being referred to as *luminous*. The radiant efficiency of a radiating source is the term given to the ratio:

$$\text{Radiant efficiency} = \frac{\text{radiant flux emitted}}{\text{power consumed}} \times 100\% \qquad (2.3)$$

Measurement of radiant flux is usually achieved by means of power-sensing transducers e.g. thermopiles. Measurement of radiant flux is termed *radiometry*.

2.8 Luminous flux, luminous efficacy and luminous efficiency

Radiant flux which contains those wavelengths which are detectable by the human eye is said to have a corresponding value of luminous flux. The unit of luminous flux is termed the *lumen*.

Theoretically the maximum attainable luminous output for unit power input is approximately 683 lumens at a wavelength of 555 nm. Thus the luminous efficacy is said to be 683 lumens per watt. The term luminous efficiency is applied to the ratio of luminous efficacy at any given wavelength to the maximum possible luminous efficacy (683 lumens per watt).

Luminous flux is usually measured by a lightmeter, containing a suitably-corrected photoelectric cell. Measurement of luminous flux is referred to as *photometry*.

2.9 Luminous intensity

Luminous intensity is a measure of the luminous flux per steradian emitted in a given direction, and is measured in candela (cd). The candela is often referred to as the luminous intensity, in a specified direction, of a light source which emits monochromatic radiation of a frequency 540×10^{12} Hz, and which has a radiant intensity in the same direction of $(1/683)$ Watts per steradian.

2.10 Illuminance and luminance

Illuminance (symbol E) is the term given to the quantity of luminous flux falling on unit area of a surface. It is measured in lux, which is equivalent to lumens per square metre. Luminance (symbol L) is the luminous intensity emitted by a light source per unit area. It is measured in candela per square metre.

Two useful examples illustrate the difference in the two parameters. Consider a petrol station forecourt as shown in Figure 2.4. The advertising sign at the entrance to the petrol station is typically lit internally by tubular fluorescent lamps. The effect is to produce a double-sided sign whose surface appears uniformly lit. The purpose of this illuminated sign is to attract the motorist to the petrol station and further to advise the motorist of the prices of the goods on offer and also of any other facilities provided. The advertising sign is not designed specifically to illuminate the area adjacent to the fuel pumps on the station forecourt. When considering the effectiveness of the advertising sign it is important to be aware of the luminance of the sign, i.e. the intensity of light emanating from the sign per unit area of the sign.

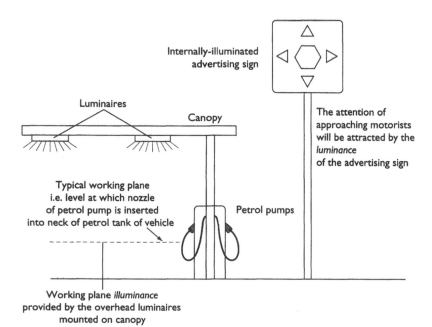

Figure 2.4 Petrol station forecourt.

Under the canopy over the fuel pumps it is usual to find luminaires installed whose prime function is to provide illumination over the work area, i.e. the level at which the motorist places the nozzle of the fuel pump into the neck of the fuel tank on the side of the vehicle. As a consequence of these luminaires, an illuminance will be produced at the working plane level. This illuminance, as measured by the simple light-meter, indicates the quantity of luminous flux falling on unit surface area of the working plane.

A second example, also involving motor transport, is that of a vehicle driver using dipped headlights during daylight. The headlights, the luminance of which can be measured, will be seen by other road users, thereby alerting them to the presence and position of the vehicle. The illuminance provided by the headlights, in the visual field of the driver of the vehicle, is often insignificant when compared to that provided by natural daylight.

2.11 Luminosity

Luminosity is often referred to as 'apparent brightness' and describes the sensation experienced by an observer who is subjected to the stimulus of luminance. Consider the example of an observer who is looking at the headlights of a car. The luminance of the headlights will be the same whether the headlights are viewed during bright daylight or in darkness. It will be evident however that the headlights appear much brighter during darkness than they do in bright daylight. In such cases the headlights have a different luminosity value in the darkness to that occurring in bright daylight.

Chapter 3

Laws of illumination

3.1 Inverse square law

Consider Figure 3.1 which represents a point source of light (S) emitting an output in the direction shown.

X, Y and Z represent three planes intercepting the light emitted from source S, such that the distance between source S and plane Y is twice the distance between source S and plane X. Furthermore the distance between source S and plane Z is three times greater than the distance between source S and plane X.

The luminous flux reaching planes X, Y and Z is constant, but the areas of planes X, Y and Z are not constant and it will be seen that the area of plane Y is four times the area of plane X and that the area of plane Z is nine times the area of plane X.

Since the same flux is illuminating a larger area, then the illuminance on the larger area must be reduced. It can be shown that the illuminance

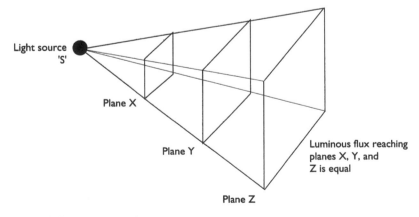

Figure 3.1 Inverse square law.

on plane Y is one quarter of that on plane X and furthermore that the illuminance on plane Z is one ninth of that on plane X.

In general terms, the inverse square law states that the illuminance on a plane varies inversely as the square of the distance of the plane from the point source.

$$\text{Illuminance (E)} \propto \frac{1}{(\text{distance})^2} \qquad (3.1)$$

It can be shown that:

$$\text{Illuminance (E)} = \frac{\text{Intensity (I)}}{(\text{distance})^2} \qquad (3.2)$$

where the illuminance is measured in lux, the intensity is measured in candela and the distance is measured in metres.

3.2 Cosine law

Consider Figure 3.2. Plane A is placed at right angles to flow of light. The plane is therefore intercepting the flow of light. The illuminance on the plane can be calculated as described in Section 3.1.

It can be shown that by rotating plane A through an angle ($\theta°$), the effect is to reduce the illuminance on the plane. The illuminance on the plane is reduced by a factor equivalent to the cosine of the angle through which the plane has been rotated.

Worked example

The cover of a telephone cable duct is located as shown in Figure 3.3. It is illuminated by a portable lamp whose luminous intensity is 750 candela in all directions. The geometric centre of the inspection cover is 2.75 metres from the downward vertical through the centre of the lamp and the height of the lamp is 2.75 metres above ground level. Calculate the value of horizontal illuminance on the geometric centre of the inspection cover.

If the inspection cover is on plane $A_1–A_2$, then the illuminance on the cover is found from:

$$E = \frac{\text{Intensity}}{\text{Distance}^2} \qquad (3.2)$$

$$E \text{ (lux)} = \frac{750}{15.125}$$

The inspection cover is, however, located horizontally and therefore the illuminance is calculated from:

$$E \text{ (lux)} = \frac{750}{15.125} \times \cos \theta$$

$$\theta = 45°$$

$$E \text{ (lux)} = \frac{750}{15.125} \times \cos 45° \qquad \text{from equation 3.3 on Figure 3.2}$$

$$E = 35 \text{ lux}$$

Thus the horizontal illuminance at the centre of the inspection cover = 35 lux.

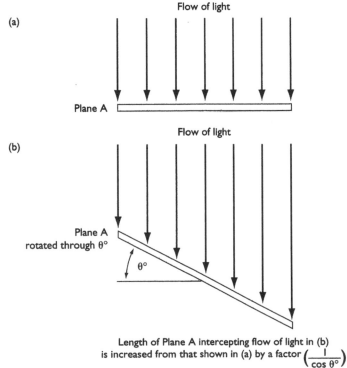

(a)

Flow of light

Plane A

(b)

Flow of light

Plane A
rotated through $\theta°$

$\theta°$

Length of Plane A intercepting flow of light in (b)
is increased from that shown in (a) by a factor $\left(\dfrac{1}{\cos \theta°}\right)$

Since luminous flux is constant,
illuminance on Plane A in (b) is reduced from that shown in (a)
by a factor $\left(\dfrac{\cos \theta°}{1}\right) = \cos \theta°$

$$\text{Thus } E = \frac{\text{Intensity}}{\text{Distance}^2} \times \cos \theta° \qquad (3.3)$$

Figure 3.2 Cosine law.

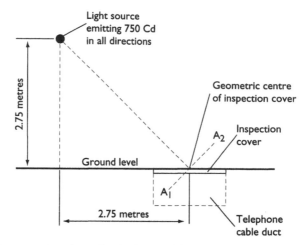

Figure 3.3 Cross-section through telephone cable duct.

3.3 Combination law

The inverse square law and the cosine law can be usefully combined into one law. Calculation of the horizontal illuminance at point A of the arrangement shown in Figure 3.4 involves trigonometry. Using trigonometry:

$$\frac{h}{d} = \cos \theta$$

from which:

$$\frac{h^2}{d^2} = \cos^2\theta$$

transposing:

$$d^2 = \frac{h^2}{\cos^2\theta}$$

Now since Illuminance (E) = {[intensity ÷ (d²)] × cosθ} it can be shown that by substituting for d²,

Horizontal
illuminance (E) $= \dfrac{\text{Intensity} \times \cos^3\theta}{h^2}$ (3.4)
(in lux) at Point A

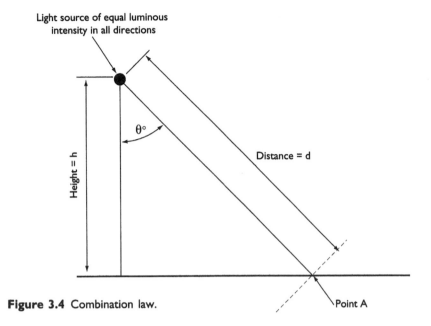

Figure 3.4 Combination law.

3.4 Relationship between luminous intensity and luminous flux

Figure 3.5 shows the relationship between luminous intensity and luminous flux. Consider a sphere of unit radius. When a cone, with its apex at the centre of the sphere, is of size such that the base of the cone,

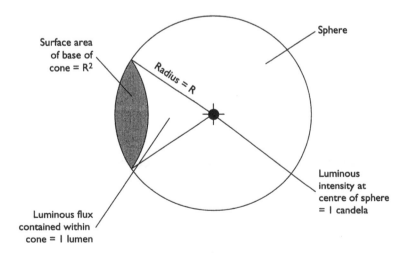

Figure 3.5 Relationship between luminous intensity and luminous flux. When radius of cone = R and surface area of base of cone = R^2, the three dimensional angle contained is termed a solid angle or a steradian. There are 4π steradians in a sphere.

on the surface of the sphere, has an area of one square unit, then the angle contained within the cone is referred to as a steradian or solid angle. The steradian is therefore a three-dimensional concept.

When the luminous intensity at the centre of the sphere is one candela, the luminous flux contained within the steradian is one lumen. It can be shown that there are 4π steradians in one sphere. In general terms:

$$\frac{\text{Intensity (I) candela}}{} = \frac{\text{Luminous flux } (\phi) \text{ in lumens}}{\text{Solid angle } (\omega) \text{ in steradians}} \qquad (3.5)$$

Worked example

Consider Figure 3.6, which refers to a beacon whose light source is emitting a beam of light in the direction shown. The light source in the beacon is a 2 kW tungsten filament lamp. The luminous efficacy of the lamp is 18 lumens per watt. If the efficiency of the luminaire optical system is 55 per cent and the solid angle of the beam of the beacon is 0.5 steradian, the intensity of the beam can be calculated from:

Luminous flux developed by the lamp	=	2000 watts × 18 lumens per watt
	=	36 000 lumens
Luminous flux emitted by the optical system	=	(36 000 × 0.55)
	=	19 800 lumens
Beam intensity	=	19 800 lumens
		───────────────
		0.5 steradian
Beam intensity	=	39 600 candela

3.5 Relationship between illuminance and luminance

The relationship between illuminance and luminance can be given by the equation:

$$\frac{\text{Luminance } (\text{cd.m}^{-2})}{} = \frac{\text{Illuminance (lux)} \times \text{Reflectance}}{\pi} \qquad (3.6)$$

where *reflectance* is defined as the ratio of reflected flux to incident flux.

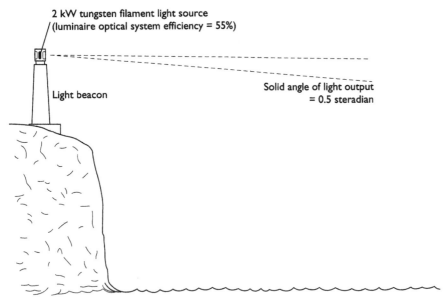

Figure 3.6 Light beacon.

3.6 Specular reflection

A specular surface is one with a mirror-like finish. When a single ray of incident light strikes such a surface, a single ray of reflected light is produced. The following laws of reflection apply:

• The incident ray, the corresponding reflected ray and the normal to the surface all lie in the same plane.
• The angle of incidence and the angle of reflectance are equal (with respect to the normal).

Figure 3.7 shows the relationships described.

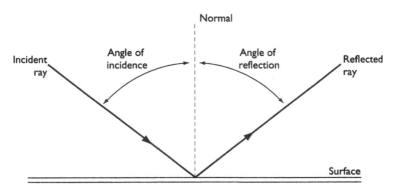

Figure 3.7 Specular reflection.

3.7 Refraction

When light passes through a boundary surface from one medium into another it will experience a change in direction. The following laws of refraction apply: the incident ray, the corresponding refracted ray and the normal to the surface all lie in the same plane. Let the medium through which the incident ray travels have a refractive index of n_1 and let the incident ray make an angle of θ_1 with the normal. Let the medium through which the refracted ray travels have a refractive index of n_2 and let the refracted ray make an angle of θ_2 with the normal.

Refractive index is defined as the ratio of the velocity of electromagnetic waves in a vacuum to the phase velocity of waves of the wavelength in the medium being considered.

It can be shown that there is a relationship linking these parameters such that:

$$\frac{n_1}{n_2} = \frac{\sin \theta_2}{\sin \theta_1} \tag{3.7}$$

Figure 3.8 shows the effects produced by refraction.

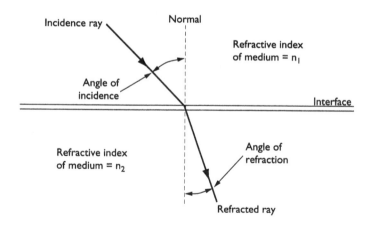

Figure 3.8 Refraction.

3.8 Dispersion

Values of refractive indices vary according to the characteristics of the medium and they also vary with wavelength of the incident light. The amount of refraction is greater for wavelengths at the violet end of the visible spectrum than it is for those at the red end of the visible spectrum.

If a sample of glass has a refractive index (μ) of say 1.5, it implies that light travels 50 per cent faster in air than it does in the glass under consideration. In general terms the velocity of light in a medium is related to the velocity of light in air by the expression:

$$\text{Velocity of light in medium} \quad = \quad \frac{\text{Velocity of light in air}}{\mu \text{ (medium)}} \qquad (3.8)$$

The shorter wavelengths of the spectrum are, with some rare exceptions, retarded by a medium more than the long waves. Thus when white light undergoes refraction, its component wavelengths are refracted to different extents and the colours become separated, resulting in dispersion.

3.9 Absorption and scattering

When a ray of light passes through a perfect vacuum it will not suffer any loss in energy. When a ray passes through a non-vacuum, or material, it will suffer energy loss due to the effects of absorption and scattering.

Absorption occurs as a consequence of processes which convert the light energy into some other form (often heat energy) although conversion to radiation, electrical energy and chemical energy also occurs in certain circumstances.

The loss in intensity from a parallel beam of light of a given wavelength, as it passes through a homogeneous medium, can be calculated from the expression:

$$I_2 \quad = \quad I_1 \cdot e^{-ax} \qquad (3.9)$$

where:

I_1 = the initial intensity of the beam of light;
I_2 = the instantaneous intensity of the beam of light after it has travelled a distance 'x' in the medium;
a = linear absorption coefficient which varies with wavelength.

It is possible, under certain circumstances, for the value of 'a' to be negative. This produces a situation where light passing through a medium will increase in intensity. This property is used to good effect in lasers.

If the medium through which light travels is heterogeneous, then scattering will occur by reflection and refraction at the boundary surfaces within the medium. In the environment, natural occurrences such as fog and cloud cause scattering in the air due to the droplets of water in suspension. Light entering a medium which is capable of scattering may be scattered back out of the material with little loss due to absorption.

3.10 Diffuse reflection and diffuse transmission

In some situations when a ray of light strikes a surface which contains inconsistencies, the light energy will spread out in all directions from the initial point of incidence, in a similar manner to that which occurs in scattering. The light which returns to the medium from whence the incident ray came is known as *diffusely reflected light*, whereas that light which passes through and into another medium is known as *diffusely transmitted light*. Figure 3.9 shows the principle of diffuse reflection.

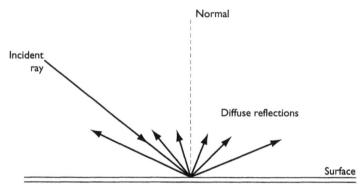

Figure 3.9 Diffuse reflection.

Chapter 4

Physiology and characteristics of vision

4.1 Introduction

The complexity of manufacturing and production techniques involved in the workplace is constantly increasing. This places correspondingly greater demands upon the human visual system.

In an attempt to strive for a safe and healthy working environment, it is essential to provide lighting whether natural, artificial or both so that:

- the workplace is illuminated to the required level;
- the lighting provided is in the correct location;
- the lighting provided does not produce detrimental effects which could lead to the development of dangerous situations.

It cannot be over-emphasized that it is not acceptable to simply consider lighting (whether supplied naturally through daylight or artificially using electric lighting) in isolation. It is essential to think simultaneously of the physiology of the human visual system, and the visual processes involved, in combination, when attempting to produce desired lighting conditions. This is a combination which is frequently overlooked.

4.2 Structure of the human eye

Figure 4.1 represents a sagittal section (cross-section) of the human eye. The eyeballs are located within the orbital cavities of the skull and are approximately spherical and of diameter typically 2.2 centimetres. More than 80 per cent of the eye (referred to as the posterior portion) lies within the skull and the remainder (referred to as the anterior portion) is exposed.

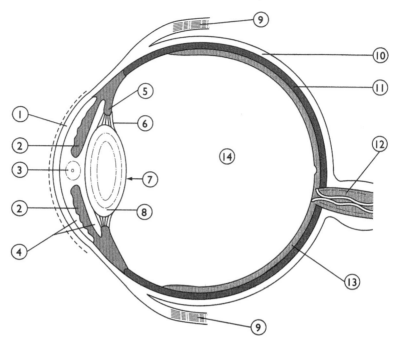

Figure 4.1 Cross-section through human eye. 1 Cornea; 2 Iris; 3 Pupil; 4 Anterior chamber containing aqueous humour; 5 Ciliary muscle; 6 Suspensory ligament; 7 Lens capsule; 8 Crystalline lens; 9 Rectus muscle (in sclera); 10 Sclera; 11 Choriod; 12 Optic nerve; 13 Retina; 14 Vitreous cavity containing vitreous humour.

The eye can be considered to consist of two separate elements:

- a light-transmitting part, referred to as the dioptric system; and
- a light-receiving part, referred to as the retina.

The inside of the eyeball consists of:

- a transparent jelly-like substance, referred to as the vitreous humour, which resides in front of the retina, in the vitreous cavity;
- a watery fluid, the aqueous humour, which occupies the anterior chamber;
- a crystalline lens, which is located in the front of the vitreous cavity;
- a diaphragm, referred to as the iris (the coloured part of the eye); and
- the ciliary muscle, which increases the curvature of the lens during close work

The eyeball consists of three coats:

- the outer coat (the sclera), which is a tough fibrous material which modifies to become the cornea, the transparent window of the eye.

- the middle coat (a vascular layer) which consists of the iris, the ciliary body and the choroid. In front of the lens is the iris. This is a ring of tissue which includes both pigment cell and smooth muscle. The pigment absorbs stray light rays. The smooth muscle is arranged in two orientations, some radially and some circumferential to the pupil. The choroid has a rich supply of blood vessels that nourish the retina. Iris blood vessels move forward to supply the anterior portion which is also nourished by the aqueous humour. The ciliary body forms a radial fringe. It is covered with pigmented epithelium and includes the ciliary muscles, which influence accommodation. The choroid is dark brown in colour and contains blood vessels and a paucity of scattered pigment cells.
- the inner coat, almost entirely consisting of the retina, is a light sensitive layer of cells. The retina can be considered as part of the nervous system and it contains two distinct types of photoreceptors, i.e. rods and cones.

The eyelids consist of two folds of tissue. Each eyelid consists of connective tissue, glands and muscle and also has rows of hairs called eyelashes. The inner face of each eyelid is covered by a mucous membrane, called the conjunctiva, which extends over the surface of the eyeball.

4.3 Photoreceptors – rods and cones

There are two distinct types of photoreceptors in the retina. These are known as rods and cones due to their physical shape. Figure 4.2 shows the structure of the rods and cones and Figure 4.3 is a schematic representation of the difference in discrimination of rods and cones. Both rods and cones contain visual pigments which are capable of absorbing incident light and converting it into small electrical potentials. In rods the visual pigment is rhodopsin or visual purple. There are three types of cones in the retina, each with a different visual pigment. Each type of cone responds differently to light of differing wavelengths, i.e. those corresponding to red, green and blue.

Rods and cones are not in equal abundance across the retina and, furthermore, they are not evenly distributed. There are approximately 120 million rods but only 5 million cones in the retina. In the central region of the retina, referred to as the fovea, there are cones only, whereas traversing towards the periphery the density of cones reduces as the number of rods increases. At the edges of the retina there are no cones at all. The distribution of rods across the retina is almost a reversal of that of the cones, with a greater density at the edges and no rods in the fovea. The threshold of stimulus of rods is lower than that of the cones and therefore colours are not detected when the prevailing light levels are low.

Most cones are connected on a one-to-one basis with an optic nerve fibre. In contrast, typically between 10 and 100 rods are usually

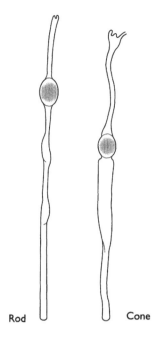

Rod   Cone

Figure 4.2 Structure of rods and cones.

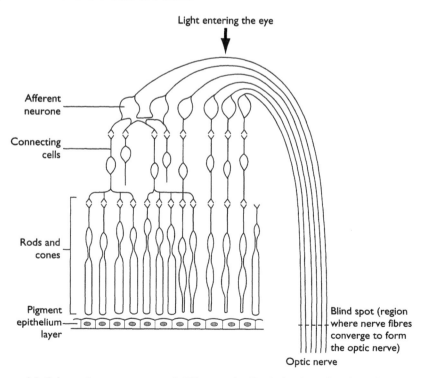

Light entering the eye

Afferent
neurone

Connecting
cells

Rods and
cones

Pigment
epithelium
layer

Blind spot (region
where nerve fibres
converge to form
the optic nerve)

Optic nerve

Figure 4.3 Schematic arrangement of difference in discrimination of rods and cones.

connected to the same optic nerve fibre. As a consequence, impulses transmitted by the rods to the brain do not necessarily originate from one fixed point on the retina.

4.4 Neurophysiology

The optic nerve can be considered as an extension of the brain and therefore is considered a part of the central nervous system (CNS). A significant amount of the function of the brain involves vision and eye movements.

When considering the functions of the brain relative to vision, the brain can be thought of as being divided into two systems, i.e. the *afferent* and the *efferent* systems. The afferent system is the *sensory* system and takes visual information from the retina and ultimately transports these signals to the brain for processing. Conversely the efferent system is the *motor* system and acts upon information that is processed by the brain insomuch as it controls eye movements.

4.5 Formation of images on the retina

In order for light entering the eye to be directed onto the retina, it must first pass through the following:

- cornea;
- anterior chamber;
- pupil;
- lens;
- vitreous humour.

So that the visual process can occur, the light reaching the rods and cones must produce an image on the retina at the back of the eye. The nerve impulses produced must then be transmitted along the visual pathway to the cerebral cortex. The development of an image on the retina is as a result of four fundamental processes:

- refraction of light rays;
- accommodation of the crystalline lens;
- constriction of the pupil;
- convergence of the eyes.

Refraction is the apparent 'bending' of light when rays pass from one medium to another one which has a different density. The light rays appear to bend at the interface between the two media. In the eye there are four such instances of refraction. Light entering the eye is refracted at the following interfaces:

- the anterior surface of the cornea ... from the lower density air to a more dense cornea;
- the posterior surface of the cornea ... into the less dense aqueous humour in the anterior chamber;
- the anterior surface of the lens ... from the aqueous humour into the more dense lens; and
- the posterior surface of the lens ... into the less dense vitreous humour.

The crystalline lens of the eye is bi-convex and additionally has the ability to rapidly change, with the aid of the ciliary muscles, its focusing power. Thus, when the eye is focusing on an object which is relatively near, the lens curves considerably in order to 'bend' the light rays and direct them towards the foveal region of the retina. This increase in curvature of the lens is referred to as accommodation, see Section 4.16.

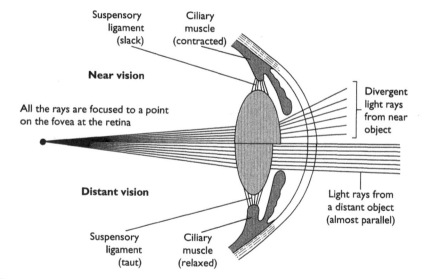

Figure 4.4 Cross-section of lens in near and distant vision.

Figure 4.4 shows a section of the lens in near and distant vision. The constriction of the iris muscles reduces the diameter of the pupil aperture through which light can enter the eye. This process occurs at the same time as the accommodation of the lens and thus it prevents light rays from entering the eye through the lens periphery. If light entered the eye through the periphery it would not be focused correctly on the retina and would lead to blurred vision and confusion.

4.6 Blind spot

In the field of vision there is a blind spot caused by the absence of photoreceptors at the optic disc. This is the point at which the optic nerve pierces the eyeball and at this point there is an absence of both rods and cones. The blind spot in each eye is located typically 15° to the lateral side of the central point of vision. Fortunately the blind spots of the two eyes are on opposing sides in their respective fields of vision. The consequence is that when the images from each eye combine into one, no part of the visual scene is left uncovered (see Figure 4.5).

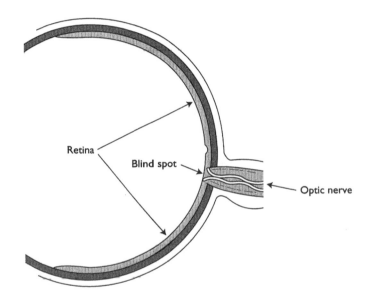

Figure 4.5 Blind spot.

4.7 Excitation of rods

Vitamin A_1 is the substance utilized by both rods and cones for synthesizing those substances which are light sensitive. Once light has passed through the eye and been absorbed into a rod, vitamin A_1 is converted into retinal which then combines with a protein in the rods known as opsin so as to form a light-sensitive chemical referred to as rhodopsin.

If the eye is not being subjected to light energy, the concentration of rhodopsin builds up to a relatively high level. If, however, the rod is exposed to light energy, some of the rhodopsin is immediately changed into lumirhodopsin. Lumirhodopsin is a very unstable substance which remains in the retina for typically less than one second. During this time it decays into another substance known as metarhodopsin, another unstable substance, which in turn divides into retinal and opsin.

Once the rhodopsin has decomposed, as a consequence of being subjected to light energy, the products of decomposition (retinal and opsin) are recombined to form new rhodopsin. During the time taken for the rhodopsin to be broken down, electrical potentials are developed. These trigger nerve impulses which travel along from the ganglion cells in the retina to the optic nerve and subsequently through the visual pathway to the brain. Figure 4.6 shows the processes involved.

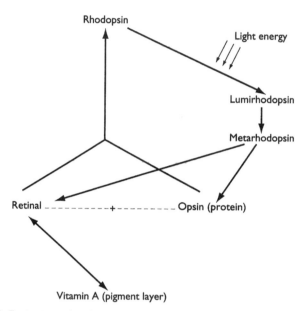

Figure 4.6 Excitation of rods.

4.8 Excitation of cones

Cones are the photoreceptors used for bright light and colour. The photopigments of the cones, unlike rhodopsin, require bright light for their breakdown and they quickly re-form. They also contain retinal but the protein portion in cones is also an opsin, but different to the opsin in rods. In cones the lipoproteins subsequently creating the photopigments are called:

- erytholabe;
- chlorolabe; and
- cyanolabe.

Only one type of photopigment is present in each of the cones. Each of the three pigments has a selective absorption characteristic dependent upon the wavelength of incident light. The characteristics overlap, and,

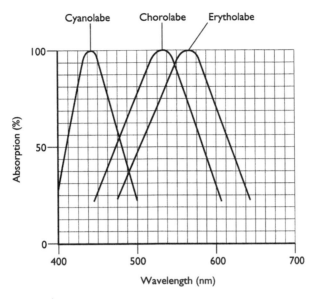

Figure 4.7 Spectral sensitivity curve for cones.

as a consequence, incident light of one particular wavelength is absorbed by more than one photopigment, but to differing extents. For example, incident light of wavelength 600 nm will be absorbed by both erytholabe and chlorolabe but not by cyanolabe.

Figure 4.7 shows the spectral sensitivity curves for the cones, from which it will be evident that the blue cones (containing cyanolabe) have a maximum absorption when the wavelength of incident light is approximately 430 nm. The corresponding wavelengths for the green cones (containing chlorolabe) and red cones (containing erytholabe) are approximately 540 nm and 575 nm respectively.

4.9 Visual pathway

The optic nerve of each eye combines at a point known as the optic chiasma, see Figure 4.8. At this point the nerve fibres carrying signals from the nasal side of the retina, cross over to the opposite side, known as the temporal side. As a consequence, information about objects on the right side of the target scene will be transported to the left side of the brain and vice versa. From the optic chiasma, nerve impulses are transmitted:

• to the brain to evoke several visual reflexes; and
• to the visual cortex in order to generate the sensation of light, colour and movement.

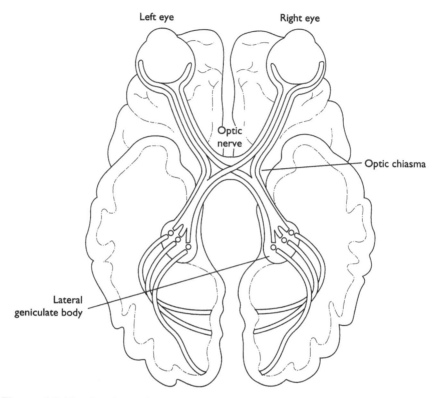

Figure 4.8 Visual pathway: horizontal section through brain.

4.10 Spectral sensitivity of the eye

The eye contains rods and cones, rods being used for low luminance, i.e. night time vision, and cones for daytime and colour vision. The relative sensitivity of rods and cones is as shown in Figure 4.9. The retina contains three different types of cone, each containing a different combination of opsin and retinal, see Section 4.8. Each of the combinations has a maximum absorption of light of a different wavelength.

Whilst appreciation of the red, blue and green (as individual colours) is relatively easy to comprehend, appreciation of how the cones detect intermediate colours, between the three primary colours, is not so straightforward. This is effectively achieved by utilizing a combination of the three cones. Thus, for example, yellow light stimulates red and green cones almost equally and in such circumstance the brain interprets this as yellow. Similarly when the red cones are stimulated between one and one-and-a-half times greater than the green cones, the brain interprets this sensation as an orange colour. In general it is the ratio of stimulation of the cones that produces the sensation of an intermediate colour in the brain.

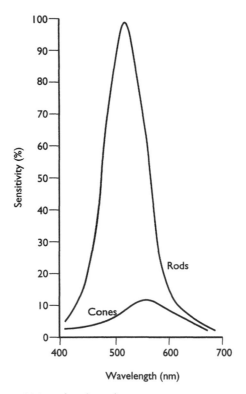

Figure 4.9 Relative sensitivity of rods and cones.

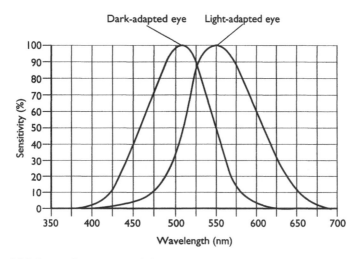

Figure 4.10 Spectral sensitivity of the eye.

The intensity of the coloured light is influenced by the strength of the signal transmitted by the cones to the brain, which is dependent upon the number of impulses transmitted by the optic nerve fibres in unit time. It is essential to appreciate that a change in intensity does not change the ratio of stimulation of the two types of cone. The ratio is the parameter upon which the brain interprets colour, this interpretation being independent from the intensity of stimulation of the cones. The spectral sensitivity of the eye is as shown in Figure 4.10.

4.11 Radiation and the eye

The prime function of the human eye is to detect visible radiation. It is important, however, to appreciate how the eye deals with both ultra-violet and infrared radiation. Exposure to radiation emitted in UV regions of the electromagnetic spectrum can lead to an individual experiencing photokeratitis.

The solar spectrum on the surface of the earth stops at wavelengths typically around 290 nm. Shorter wavelengths are eliminated by the ozone layer in the atmosphere. Short wave rays are capable of destroying both organic substances and living organisms. Such light can cause blindness if the eye is exposed excessively.

It is equally important to appreciate the effect of infrared radiation on the eye. IRB in the range 1400 nm to 1900 nm is almost totally absorbed by the combination of the cornea and the aqueous humour and radiation greater than the 1900 nm is absorbed by the cornea. IRB and IRC radiation which is absorbed by the ocular media, or in the cornea, will produce a temperature increase in these tissues, which may lead to opacities such as cataract.

4.12 Visual threshold

Visual threshold is the minimum amount of light energy required to evoke the visual sensation. A large amount of light entering the eye is absorbed by the retina and any stray light is absorbed by the iris and choroid, which prevents light being reflected back into the retina which would cause confusion. In visual sensation, the number of quanta of light actually absorbed by the photosensitive pigments is significant.

In the dark-adapted eye, the peak sensitivity occurs, in scotopic vision, at a wavelength of 505 nm compared to a peak sensitivity for the light-adapted eye of 555 nm. Thereafter the level of visual threshold is influenced by:

- the wavelength of the light stimulus;
- the state of adaptation of the eye;
- that part of the retina which is stimulated;

- the size of the light stimulus; and
- the duration of the light stimulus.

Whilst the value of visual threshold will differ from individual to individual, it is typical to refer to a value that will have a 50 per cent chance of evoking a response in 50 per cent of any population.

4.13 Photopic and scotopic vision

The retina is a sense organ with a dual function, i.e. rod or scotopic vision which is without colour, and cone or photopic vision which is accompanied by colour. There is, however, no clearly defined demarcation line as to where rods cease to operate and cones start to operate, and vice versa. The two systems will operate simultaneously over intermediate ranges of light intensity.

Visual sensitivity, which is the term applied to the ability to see and clearly distinguish objects, is influenced by the concentration of visual pigment and so with a constant level of illuminance the visual pigments will also attain a steady-state operating value.

It is an everyday experience for an individual to be temporarily visually affected when moving from a light environment into a dark environment, and vice versa, and a time duration is required before the visual system will adapt to the new surroundings. This time is called *adaptation time*. Adaptation will be considered in detail in Section 4.18.

Visual pigments do not absorb equal amounts of light of different wavelengths. Since the amount of light absorbed influences the smallest quantity of light energy required to produce a visual sensation, referred to as the *visual stimulus level*, the absorption curve should approximate to a visibility curve.

Whilst scotopic vision is to all intents and purposes colourless, the rods are not equally sensitive to the different spectral colours. The scotopic visibility curve shows that the dark-adapted eye is most sensitive to light of approximate wavelength 505 nm. The scotopic visibility curve coincides with the absorption curve of rhodopsin, the visual pigment present in rods.

Scotopic vision, i.e. that occurring in relatively low prevailing luminance levels, exhibits a strange characteristic. When viewing an object, it is less readily seen when it is made the centre of visual attention than it is when it is detected out of the corner of the eye. It follows therefore that people who are involved as lookouts during the hours of darkness will have to scan the night scene slowly rather than concentrate the gaze on definite outlines, since peripheral detection is far more accurate than foveal detection. Stars in the night sky are more easily detected when looking to one side of the star.

In contrast, in photopic vision, which relies on cones, the light adapted eye exhibits a maximum sensitivity at a wavelength of approximately 555 nm. The change between sensitivities of the light- and dark-adapted eyes is referred to as the *Purkinjé shift*.

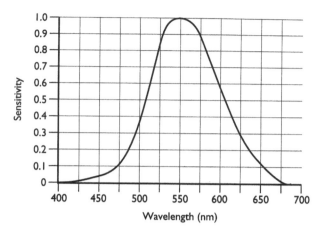

Figure 4.11 Spectral luminous efficiency of the light-adapted eye (V_λ).

If the y-axis of the light-adapted eye characteristic shown in Figure 4.10 is converted to a sensitivity scale of maximum value of 1.0, then the plot is referred to as the *spectral luminous efficiency for photopic vision* (V_λ). Figure 4.11 shows this characteristic, from where it can be seen that the relative spectral luminous efficiency of the light-adapted eye at a wavelength of 555 nm is 1.0 with a corresponding value of approximately 0.2 when the wavelength of incident light is 635 nm.

Photopic vision corresponds to a luminance greater than 3 candela per square metre whereas scotopic vision corresponds to a luminance less than 0.0003 candela per square metre.

4.14 Mesopic vision

Mesopic vision is the term given to vision lying between photopic and scotopic vision. When the prevailing luminance of a scene is increased from that applying under scotopic vision conditions, three effects can be observed:

- foveal detection and peripheral detection move towards becoming equally attainable;
- colour sensation appears increasingly stronger as the luminance increases; and
- there is a change in the relative luminosity of objects of different colours.

It will be evident that as the luminance level increases from that under scotopic conditions to that under photopic conditions, the overall response of the eye, to incident light of different wavelengths, will lie between the two curves shown in Figure 4.10. Mesopic vision corresponds to luminance values between and including 0.0003 candela per square metre and 3 candela per square metre.

4.15 Visual discrimination

This is the term given to the ability of the retina to distinguish between certain visual stimuli. There are three separate types of visual discrimination:

- light discrimination;
- spatial discrimination; and
- temporal discrimination.

Light discrimination is the term given to the ability of the eye to detect a very weak light source. When assessing an individual's visual field, he or she is asked to fixate the eyes straight ahead on a given object of interest whilst relatively weak lights are simultaneously flashed in the periphery. When the individual detects the flashing light, his or her response is recorded by striking a button. The light intensity is subsequently reduced until the individual's visual threshold is attained. The light sensitivity is also correlated with colour vision such that reduction in the illuminance and hence luminance levels will produce a corresponding reduction in colour perception.

Spatial discrimination is the ability of an individual's visual system to recognize the shape of objects, e.g. a pattern or a letter of the alphabet. Additionally, it can be considered as the ability to resolve separate parts of the same pattern. Thus M∧T is recognized as the word MAT despite the symbol ∧ being incomplete. Spatial discrimination can therefore be considered as a form of *recognition*.

Temporal discrimination is the term given to the ability of an individual's visual system to detect sensations produced as a consequence of time-varying stimuli, e.g. flickering lights. The flickering of the picture on a television screen, by way of example, is so fast that the visual system detects this as one continuous picture.

4.16 Accommodation, convergence and stereopsis

Binocular vision is the simultaneous vision with both eyes when they fixate on a particular object of interest. The images from both eyes are combined to produce one composite image. This is referred to as *binocular single vision* and it is considered in conjunction with *stereopsis*, which is the binocular perception of depth, when both eyes view an object of interest from slightly different angles.

The two eyes are separated slightly on the horizontal plane and so therefore the two images formed on the retina will differ slightly. This difference in images will provide enough visual information to enable an estimation of depth to be made.

If an individual suffers a total loss of vision in one eye, they will also suffer from problems in judging depth, and whilst monocular cues will reduce the problem allowing some depth perception, stereopsis will be lost permanently.

The effect of loss of stereoscopic vision can be simply but effectively illustrated when an individual tries to thread a needle with one eye closed or tries to pour a cup of tea whilst closing one eye. Furthermore the role of binocular vision can be demonstrated when using a stereo-scope. This is a device which enables a viewer to simultaneously look at two photographs of the same scene but taken from differing angles. This difference corresponds to spatial separation between the two eyes. One eye sees one picture and the other eye sees the other picture. The resultant effect is a single image which takes on a three-dimensional appearance.

Monocular cues are cues that an individual can apply to a particular visual task to aid in the judging of distance. Such cues are termed monocular since they can be interpreted by one eye just as effectively as by both eyes. Examples of monocular cues include:

- aerial perspective;
- interposition;
- linear and textural perspective;
- movement parallax;
- relative size of objects;
- shadowing; and
- upward dislocation.

When an individual looks into the distance, the light rays entering the eye will run almost parallel to the optic axis. The ciliary body is relaxed and the lens fibres are pulled tight, maintaining the lens as flat as possible.

If the object of interest becomes nearer to the individual, a situation will develop whereby at a certain distance between the individual and the object, light rays will enter the eye at such a large angle in relation to the optic axis that the image subsequently projected onto the retina will lose focus and appear to be blurred. In order to prevent this from occurring, the radius of curvature of the lens increases, which is achieved by the contraction of the ciliary muscle. When the curvature of the lens takes on the larger radius, refraction of the incoming light increases and the visual image will remain in focus.

As the object of interest becomes closer to the individual, further contraction of the ciliary muscle enables the lens to become correspondingly more convex. This process is referred to as *accommodation*. The reverse process is also true, there is a relaxation of the ciliary muscle resulting in a decrease in the radius of curvature of the lens in order to focus on distant objects. This is a *relaxation of accommodation*.

If a visual task is placed close to an individual's eyes, a point is reached at which the object of interest becomes blurred. This is the point at which the lens of the eye has attained its natural maximum convexity. This point, at which the refraction of light reaches its maximum and at which the object of interest is just in focus on the retina, is referred to as the *near point*. The position of the near point is significantly influenced by the flexibility of the lens. In newborn babies, the near point can be as close as one centimetre from the eye whereas as individuals grow older

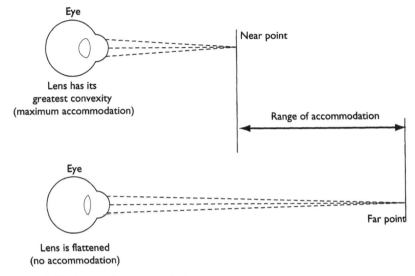

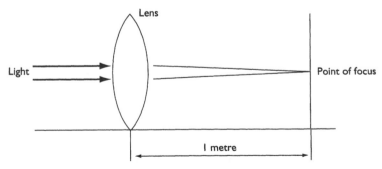

Figure 4.12 Range of accommodation.

the natural maximum convexity progressively reduces, hence the need for reading spectacles with increasing age.

At the other end of the visual range is the *far point*, at which adjustment of the radius of curvature of the lens need only be minimal in order to produce a distinct visual image on the retina. The far point in a normal sighted individual is typically infinity although in an eye examination 6 metres is equated to infinity.

The distance between the near point and the far point is referred to as the *range of accommodation* and is shown in Figure 4.12.

The refractive capabilities of a lens are expressed in units of *dioptres* (symbol D). The dioptre can be defined as the strength required of a lens in order to focus parallel rays of light at a distance of one metre. Thus a one dioptre lens will have a focal length of one metre (see Figure 4.13).

Figure 4.13 The dioptre. Lens power = 1 dioptre or 1 D.

The *dioptric power* of a lens can also be expressed as the inverse of the focal length. Thus for example for a 10 D lens, the focal length is (1/10) = 0.1 metre = 10 centimetres.

The combined refractive strength of the eye, when at rest, is approximately 60 D. The breakdown of this combined approximation is given as:

- 48 D cornea and intraocular fluid;
- 12 D flattened lens; and
- < 1 D vitreous fluid.

Typically an additional 14 D can be achieved by accommodation of the lens in a young individual. The ability to accommodate decreases with age as the lens hardens.

When an individual stares into the distance both eyes will gaze straight ahead and the visual axes will be parallel. If the gaze is centred upon a closer visual task, both eyes will rotate towards the mid line. This process is referred to as convergence.

Convergence is achieved by contractions of the medial rectus eye muscles. The opposite, *divergence*, occurs when objects of interest move away from the individual. The lateral rectus muscles (one each eye) diverge the eyes. The visual axes will become more parallel due to the contraction of the lateral rectus eye muscles.

4.17 Field of vision

The visual field for each eye is typically:

- laterally 85°
- up 45°
- nasally 60°
- down 65°

Figure 4.14 shows the typical visual field. The visual field can become affected by diseases of the eye such as primary open angle glaucoma.

4.18 Adaptation

The effect of a visual stimulus is not constant since it is influenced by the state of adaptation of the retina at the time a stimulus is introduced.

4.18.1 Dark adaptation

When an individual enters a darkened interior from a lighter one, the quantity of rhodopsin in the rods is initially very small and the

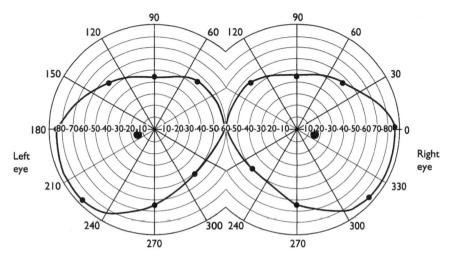

Figure 4.14 Normal field of vision.

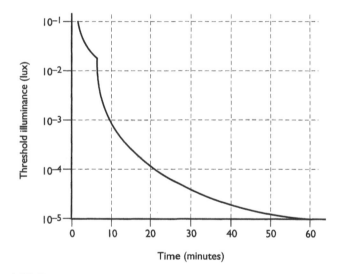

Figure 4.15 Dark adaptation process.

individual can see very little. Because the amount of light energy in the darkened interior is also very small, little of the rhodopsin formed in the rods is broken down. As a consequence, the concentration of rhodopsin accumulates over a period of time and ultimately attains a value sufficiently high for light to stimulate the rods. On moving from a light environment into a dark one, dark adaptation occurs, the retina sensitivity increases relatively quickly during the first phase (typically ten minutes) and thereafter more slowly. Figure 4.15 shows a graphical representation of the dark adaptation process.

4.18.2 Light adaptation

When an individual passes from a dark environment into a light one, the process of light adaptation will occur. When large amounts of light energy strike the rods, large amounts of rhodopsin are broken down. Since the formation of rhodopsin is a relatively slow process, the concentration of rhodopsin in the rods falls to a relatively low value. Similar effects occur in the cones and the sensitivity of the retina greatly reduces. The sensitivity of the retina decreases as the light energy increases.

Figure 4.16 shows the variation in sensitivity of the retina during the light and dark adaptation processes.

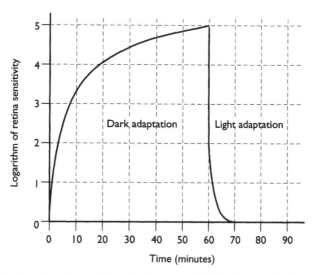

Figure 4.16 Variation in sensitivity of the retina during dark and light adaptation processes.

4.18.3 Problems associated with twilight

It is usual for the transition between day and night vision to take place slowly whilst the illuminance level is reducing. During this transition a point will be reached, before the transition is complete, when both rods and cones are functioning simultaneously. It will be evident that at this point both the rods and cones are not functioning at their optimum efficiency values. This situation is one which affects motorists, for example, when driving at twilight. Whether the vehicle driver is looking straight ahead (and therefore relying on cones) or using peripheral vision (relying on rods), the image can be unclear.

4.18.4 Colour adaptation

If the eye is exposed to a powerful light for several minutes, its sensitivity will be suppressed. It follows that the colours of the objects in the field of view appear different to an eye which has been exposed to the offending colour than they do to a normal eye. This phenomenon is referred to as colour adaptation.

4.19 Visual acuity

Visual acuity is the term given to the ability of the eye to discern detail. Visual acuity varies considerably over the retina, from very fine acuity at the fovea, where there are cones only, to very poor acuity at the periphery, where rods prevail.

The value of prevailing illuminance, and hence luminance, will influence the visual acuity up to a point beyond which additional illumination will have no beneficial effect. Figure 4.17 shows a typical relationship between visual acuity and prevailing illuminance. The point at which the characteristic levels off is dependent upon the degree of difficulty of the visual task itself. It will, of course, vary from task to task.

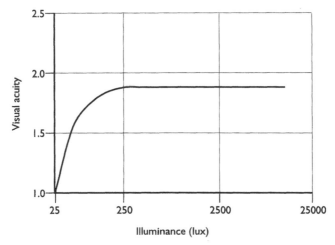

Figure 4.17 Typical relationship between visual acuity and prevailing illuminance.

Measurement of visual acuity can be achieved in several ways:

- It can be expressed as the reciprocal of the angle subtended at the nodal point of the eye at a distance where the extremities are detectable as separate entities (see Figure 4.18). Using this method, the resolution for the normal eye is approximately 1.0, which means that

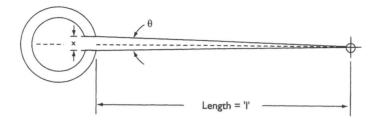

Visual acuity = $(\frac{1}{\theta})$ where θ is measured in minutes of arc (one minute = $(\frac{1}{60})$ x degrees)

Figure 4.18 Measurement of visual acuity.

a person with 'normal' vision can distinguish two separate points when the visual angle is one minute of arc. An individual with 'normal' vision, where an object placed at a distance of 6 metres or greater, forms a clear image on the retina of the unaccommodated eye, is said to be *emmetropic*.

By way of example, the standard of static visual acuity in the UK for ordinary (non-professional and vocational) vehicle drivers is the ability to read in good light, with the aid of glasses or contact lenses if worn, a vehicle registration mark affixed to a motor vehicle when the height of the characters is 79.4 mm and the distance between the observer and the registration mark is 20.5 metres. Details of visual acuity requirements for various occupations can be found in the Association of Optometrists Handbook.[1]

When the object of interest is moved further away from the eye and so therefore its visual angle becomes progressively smaller, detail on the object becomes more difficult to discern and ultimately becomes imperceptible.

- Using a Snellen chart. The individual reads decreasing sizes of printed letters when positioned 6 metres from the chart. This method effectively compares the individual with one whose vision is measured as 'normal'. Thus, if an individual is positioned 6 metres away from the chart and at this distance he or she can only clearly detect that line of letters on the chart which can be clearly detected at a distance of 12 metres by a person with 'normal' vision, then the individual is said to have 6/12 vision. Figure 4.19 is a typical Snellen chart. In practice, a separation distance of 6 metres between the patient and the Snellen chart might not be practical. One method of overcoming this problem is to use a chart with the letters reversed, and to ask the patient, who is seated 3 metres away from the chart, to read the letters through a plane mirror so that the effective distance between the patient and the chart is 6 metres. *Distant vision* is estimated using the Snellen chart.
- Black Landolt rings, with a gap equivalent to the thickness of the ring, can be used. The gap should also be 20 per cent of the diameter of the ring. It is usual to have a grid of such rings and to randomly orientate the gaps.

Figure 4.19 Typical Snellen chart.

- *Near vision* is tested by the individual reading words composed of letters of different sizes and near visual acuity is a measure of the ability to read words composed of letters of different sizes at a distance of 33 cm. The text is based upon the 'N' type point system of measurement of the height of letters used in printing, where one point is equal to 1/72 of an inch. Examples of the sizes of 'N' types are shown in Figure 4.20.

4.19.1 Photopic acuity

The maximum resolution of the eye in foveal vision and when operating in bright light is typically seven times greater than that obtained in full moonlight. This corresponds to a visual angle of approximately 0.4 minutes of arc, which is close to the angular separation of the cones in the retina.

4.19.2 Relationship between cones and visual acuity

Individuals normally have very acute vision in the central region of the visual field. This is because in the fovea, which is in the centre of the

Size N.8

Police, fire, ambulance, blood transfusion service, mines rescue, mountain rescue, lifeboat services and bomb disposal services may use blue flashing lights fixed to vehicles.

Size N.12

Police, fire, ambulance, blood transfusion service, mines rescue, mountain rescue, lifeboat services and bomb disposal services may use blue flashing lights fixed to vehicles.

Size N.24

Police, fire, ambulance, blood transfusion service, mines rescue, mountain rescue, lifeboat services and bomb disposal services may use blue flashing lights fixed to vehicles.

Figure 4.20 Examples of sizes of N-type letters.

retina and of typically 1 mm in diameter, there are no rods present. In this region the cones are smaller in diameter and much more densely packed than those in the peripheral region of the retina. Furthermore, in the foveal region each cone connects through an almost direct pathway to the brain, on a one-to-one basis, i.e. one cone to one optic fibre. The result is that impulses generated from each cone do not become confused with impulse from other cones.

4.19.3 Scotopic acuity

Acuity of rod vision is strongly influenced by the level of illumination. Thus, for example, with a flash of light there will be a corresponding random distribution of light quanta absorbed and it will be evident that the more intense the flash of light, the more dense the distribution. Under low illumination conditions, an object of smaller diameter than the mean distance between the light quanta will remain undetected and the accuracy of vision in such circumstances is determined by the pattern of quantum excitation of the light-sensitive retina.

4.19.4 Static and dynamic visual acuity

Reading a fixed chart, e.g. a Snellen chart, is a measure of an individual's static visual acuity, since the object of interest clearly does not move. Dynamic visual acuity, which is often considered as a more useful measure of acuity (since it relates to the real world), refers to the acuity of a moving target. A typical everyday example of dynamic visual acuity is the driving of a car, where the driver (whilst in a moving vehicle) is constantly looking for gaps which may, for example, be side road entrances or spaces between travelling vehicles when considering an overtaking manoeuvre.

4.20 Visual perception

The impulses generated from the nerves, as a consequence of light falling onto the retina, enter the visual cortex and are subsequently interpreted. The brain constructs recognizable images from the multitude of potentials generated by the cells in the retina. This process is known as visual perception. In the perception process the images on the retina, which are reduced in size, inverted and two-dimensional are re-inverted, enlarged and furthermore made to appear as three-dimensional by activities in the brain.

Figure 4.21 Perception by sight.

The process in which images are understood is linked with inherent and learned features. For example, if an individual sees a ferocious dog there is an inherent impulse to move quickly away. Similarly when in a restaurant, it is natural for an individual's mouth to water when a favourite meal is placed on the table. It should be appreciated that the same meal may induce a feeling of malaise in another individual who recalls not liking that particular food.

4.20.1 Perception by sight

It is evident that the process of seeing is far more involved than just simply the projection of an image on the retina of the eye. The interpretation of what an individual sees can vary. Consider Figure 4.21, from which it will be apparent that an individual's interpretation can change. On one hand it is possible to visualize an ornamental vase whilst on the other hand the white colour may appear in the background and an individual will visualize the profiles of two heads facing each other.

Chapter 5

Defects and anomalies of vision

5.1 Introduction

It is beyond the scope of this book to provide an in-depth appraisal of all possible defects of the human visual system. It is important, however, to consider some of those visual defects which may lead to problems within the workplace.

5.2 Glare

In situations where there is excessively high luminance contrast, glare will develop. It is one of the major factors in the visual environment. The presence of a high luminance source of light in the field of view will react unfavourably upon the visual system in two ways:

- the ability of the eye to discern detail in the vicinity of the high luminance source will be reduced, termed *disability glare*;
- there will be a *discomfort glare*, the effects of which can vary between a mild brightness distracting attention and an intolerable painful sensation.

5.2.1 Induction and disability glare

The influence of one region of a visual field on another part of the visual field is referred to as induction. Reference to Figure 5.1 will reveal an apparent 'lightness' difference in the two halves of the ring, the effect being more pronounced if a dividing line is placed across the boundary between the two halves. This phenomenon is an example of *monochromatic induction* where stimulation of part of the retina reduces the sensation provided by stimuli in adjacent areas of the retina.

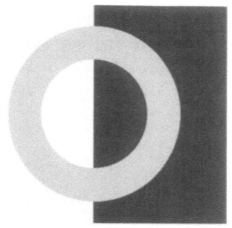

Figure 5.1 Induction and disability glare.

In addition there are inductive colour effects. If the two halves of the ring shown in Figure 5.1 were replaced by different colours, the two halves would still appear different and would have tinges of the complementary colours to their corresponding backgrounds. When induction occurs, there is an accompanying reduction in foveal visual capability as a consequence of unwanted light stimuli in the periphery of the retina.

The cause of impaired visibility in the vicinity of a high luminance source, and hence a glare source, can be understood when it is recalled that the instantaneous visual range of an individual is determined by the adaptation level of the eye. In a low-luminance visual scene, which is unaffected by glare, the eye adaptation allows detection of small differences in luminance across the field of view. However if a high luminance source, a glare source, is suddenly introduced into the field of view, the adaptation level of the eye will be raised, more so in the retina area of the image of the source of glare, and as a consequence the smallest detectable difference in luminance will be higher than that applying before the introduction of the glare source. It will be evident that much of the field of view which could previously be seen clearly will now become invisible.

5.2.2 Phototropism and discomfort glare

The tendency of plant life to grow in the direction of light is a well-documented phenomenon and is referred to as phototropism. The eye is similarly drawn towards light, in the absence of other controlling actions, and will fixate on bright regions in the field of view. Luminaires in an interior are a source of high luminance, and, unless measures are taken to reduce the effects, they can lead to discomfort glare in individuals who operate in their vicinity. There is an element of phototropism in discomfort glare.

5.3 Persistence of vision

Whilst the eye is often compared to the simple camera, there is one very distinct difference between them, which relates to the persistence of vision. The temporal summation in the eye does not extend for time exposures typically greater than 0.1 second. The eye takes an appreciable time to become aware of a target object in the visual field, typically one tenth of a second with a medium illumination, and, further, that vision will persist for also approximately one tenth of a second after the target object has disappeared from view. This phenomenon is referred to as persistence of vision. It is of course this property of the eye that allows us to obtain a smooth and continuous appearance from a cinema film.

5.4 After-images

If an individual looks for a short time (typically one or two seconds) at a tungsten filament lamp and then closes their eyes, a bright image of the filament lamp will be experienced, referred to as a *positive after-image*.

If an individual looks at a powerful red lamp for a relatively short time duration and then transfers the gaze to a well-illuminated sheet of white paper, a blue-green image of the lamp will be experienced. This is referred to as a *negative after-image* and its colour is complementary to that of the original stimulus. The sensitivity of the red receptors, in that area of the retina where the original red image was developed, is suppressed, and therefore the white paper stimulates blue and green receptors, ultimately giving rise to a peacock blue image.

5.5 Flicker

If the retina is stimulated by flashes of intermittent light at a particular rate of stimulation, a sensation of flicker will be produced. If the rate of flashes is greater than a certain value, the previously perceived discontinuous sensation will become continuous. The value of frequency at which the flicker becomes an apparent continuous signal is referred to as the *critical fusion frequency (CFF)*.

The effect is attributed, inter alia, to the combination of each flash of incident light falling upon the retina together with the positive afterimage of the preceding stimulus. The value of CFF is known to vary with:

- light intensity;
- size of stimulus; and
- wavelength of incident light.

The approximate range of CFF values is between 15 flashes per second with low intensity stimulation and 60 flashes per second with high intensity stimulation.

5.6 Hypermetropia

This is a visual defect where seeing is clearer at greater distances. When the eye is relaxed, the total powers (in combination) of its optical elements is insufficient to bring visual targets into focus hence light focuses behind the plane of the retina, see Figure 5.2. An individual who is far-sighted, i.e. hypermetropic, may well suffer from fatigue and discomfort when attempting to focus, particularly with close work where the visual task is more demanding.

Symptoms of far-sightedness (or hypermetropia) include:

• difficulty in maintaining clear vision during reading;
• difficulty in concentrating; and
• visual fatigue following close work.

The solution for hypermetropia is for the patient to wear either plus-powered (convex lens) spectacles or contact lenses.

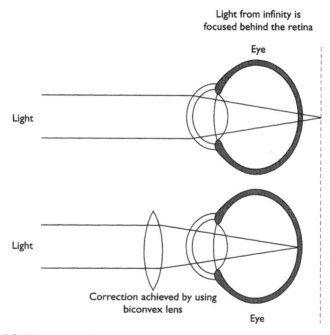

Figure 5.2 Hypermetropia.

5.7 Myopia

Myopia, or near-sightedness, is a type of visual defect where seeing at closer distances is easier than seeing at far distances as the image is

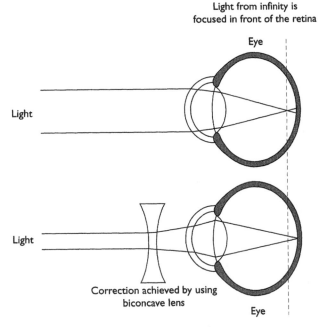

Light from infinity is
focused in front of the retina

Eye

Light

Light

Correction achieved by using
biconcave lens

Eye

Figure 5.3 Myopia.

produced before the plane of the retina, as shown in Figure 5.3. If an individual is near-sighted, the near point of clear vision can in some cases be very close to the eye, typically 10 centimetres.

Simple myopia does not always result in discomfort since the eye is not called upon to make changes in adaptation as is the case with hypermetropia. In near-sightedness, the image of the visual target falls in front of the retina and is therefore out of focus. The solution is for the sufferer to wear either negative-powered (concave lens) spectacles or contact lenses (see Figure 5.3).

5.8 Presbyopia

During the ageing process, an individual loses the ability to focus clearly at the normal reading distance. Such a situation is referred to as presbyopia. There is no precise age at which this occurs, it differs from individual to individual, but in general terms it often starts to occur between the ages of forty and fifty. It is thought to be caused by two factors:

- hardening of the crystalline lens making it difficult for the ciliary muscle to alter its shape; and
- weakening of the ciliary muscle.

Presbyopia occurs in addition to any existing vision defects and the usual symptoms associated with presbyopia include:

- blurred vision at the usual reading distance;
- a necessity to hold reading material progressively further away;
- headaches;
- visual fatigue when performing near tasks; and
- difficulty in concentrating when carrying out close detailed work.

It is not possible to prevent presbyopia but it can be compensated for with reading spectacles or bifocal contact lenses.

5.9 Astigmatism

Astigmatism is the condition where an individual will simultaneously interpret equal length radial lines as having different lengths. It is often as a consequence of unequal curvatures of the cornea or lens in different planes, and, therefore, if the lens of the eye is focused correctly for light entering the eye at one particular angle, it well be found that the lens will be incorrectly focused for light entering the eye at a different angle. Figure 5.4 shows one method used to test for astigmatism.

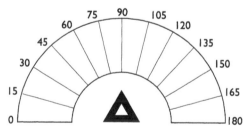

Figure 5.4 Fan line test for astigmatism. An individual with astigmatism will see some lines of the fan as distinctly black and other lines as grey. This information will allow an optometrist to orientate a cylindrical lens in order to correct the fault

5.10 Diplopia

Good vision will of necessity require the ability of both eyes to aim at the same target simultaneously. Essentially both eyes work as a team, and, if this is not achieved, double vision (diplopia) will exist or the brain will attempt to block out the image of one eye (suppression). In many of the cases of double vision, the cause is as a consequence of some malfunction in the muscle or its nervous control. Each of the two eyes has six extraocular muscles, and, if they are not all functioning correctly, improper alignment will occur.

Double vision, as a consequence of faulty alignment, is often very uncomfortable and the brain will attempt to block off one image, of the two presented, in order to restore some degree of visual comfort. In some cases extra muscular effort will be required in order to maintain correct alignment which can lead to headaches and general visual discomfort.

5.11 Colour blindness

The ability of the eye to distinguish different colours is dependent heavily on the behaviour and performance of the cones. Any malfunction of the cones can lead to abnormal colour appreciation, many forms of which are congenital in origin.

5.11.1 Monochromats

The monochromat exhibits a total absence of colour appreciation and so the individual is totally colour blind (monochromatism).

5.11.2 Dichromats

A dichromat is an individual who shows an absence, but not an anomaly, of one factor in the cones, and this can take one of three forms:

- a protanope when the red factor is absent;
- a deuteranope when the green factor is absent; and
- a tritanope when the blue factor is absent.

5.11.3 Trichromats

The trichromat shows an anomaly, but not an absence, of one factor in the cones. This also may take on one of three forms:

- a protanomalous when the red factor is weak;
- a deuteranomalous when the green factor is weak; and
- a tritanomalous when the blue factor is weak.

Anomalous trichromats exhibit a colour difference rather than a colour blindness.

5.12 Colour vision tests

Colour vision tests should be carried out ideally in daylight or alternatively under an illuminant which does not produce unwanted colour distortion. The test used include:

- **Ishihara test:** This test uses a pattern of dots of different sizes which in combination construct a symbol against a background. By careful selection of the colours of the dots, the symbol will be either invisible or different to individuals with anomalous colour vision.
- **The 100 hue test:** This test involves 85 coloured caps which the individual is asked to arrange in order of 'just-noticeable' difference of hue. It will be evident that only those individuals with exceptional colour vision will be able to rank the 85 caps in correct order.
- **City University test:** This test consists of ten plates, each of which shows five spots, i.e. one central spot and four radial spots of differing colours. The individual whose colour vision is being assessed has to select which of the radial spots correctly matches the colour of the central spot. The colour of the radial spots is carefully selected so that the test reveals those who exhibit loss of red, green and blue perception respectively.
- **The anomaloscope:** This test requires individuals to match a standard yellow colour against that of a red/green mixture. The proportions of red and green in the mixture required by the individual in order to obtain a match will indicate to the optometrist the degree of colour deficiency shown by the individual.
- **Lantern tests:** These tests require the individual to assign a name to the colours perceived. Signal lanterns with different coloured lights are simulated and the individual is asked to name the colours shown. These tests are frequently used in the Navy and other seafaring professions.

5.13 Night blindness

This refers to an individual whose vision is considered as normal under conditions of good illumination but is considered as defective under dimly-lit conditions. The rods contain a visual purple which is bleached by light. Visual purple is formed from vitamin A_1 and so an individual with a deficiency in vitamin A_1 will lack visual purple and will suffer from night blindness. It will be evident that individuals who require good night time vision in order to carry out particular visual tasks, e.g. airline pilots, must have sufficient vitamin A_1 in their diet.

The glare a motorist experiences from the headlights of an oncoming vehicle rapidly breaks down rhodopsin to such a level that at a speed of 80 kilometres per hour an average driver will travel 'blind' for a distance ranging typically between 20 and 25 metres. If however the same vehicle driver is deficient in vitamin A_1 this short-term blindness can be increased by a factor of three.

There is, however, an optimum value of vitamin A_1 required by the retina, and, contrary to 'old wives' tales', excessive intake of the vitamin will not lead to the development of exceptional sight. Night blindness may be present in children who have suffered severely with measles or malnutrition.

5.14 Effects of age

Changes in the ageing eye can be divided into 'perceptual' and 'physical'. Perceptual changes include:

- reduction in visual acuity (both static and dynamic);
- reduction in contrast sensitivity;
- reduction in peripheral vision;
- reduction in dark adaptation time;
- greater scattering of incoming light causing glare; and
- reduced accuracy in identifying some colours, especially blue.

Physical changes include:

- presbyopia
- changes in proteins of lens due to absorbing UV radiation; and
- changes in vitreous humour.

In general, older workers are doubly affected in working environments where there is weak contrast and weak luminosity. Initially they require more light in order to be able to see an object, but simultaneously they are disadvantaged in that, due to the increased luminance which accrues, they become more quickly dazzled by sources of glare.

The visual discomfort experienced by older workers is increased when they change rapidly from areas which are well lit to areas which are poorly lit.

5.15 Glaucoma

The human eye is filled with fluid which is produced and drained off at similar rates. If the rate of fluid production is greater than the rate of drainage, the fluid will build up and the subsequent pressure in the eye will increase. In early stages, the visual defects so produced are not readily detected by the patient. However over a period of time and with the intraocular pressure (IOP) remaining high, the optic nerve atrophies further, producing extensive visual field defects which may cause workers so affected to have higher than usual accident rates.

Chapter 6

Colour

6.1 Colour definitions and terminology

Colour can be characterized by three parameters, i.e. hue, lightness and saturation, sometimes referred to as colour, value and chroma respectively.

6.1.1 Hue, lightness and saturation

In order to be able to distinguish between reds, blues and yellows for example, it is usual to refer to hue. It is also possible to refer to colours as light and dark and we refer to the lightness of colours. The dullness or vividness of colours can be referred to as saturation.

6.1.2 Munsell colour solid

Colour parameters can be conveniently represented on a three-dimensional solid figure known as the Munsell colour solid. The colour solid, which resembles a colour wheel, is shown in Figure 6.1, which is reproduced by kind permission of Minolta (UK) Limited.[2]

Using the colour solid, it is possible to specify any individual colour by a sequence of letters and numbers. The saturated colours are distributed around the 'equator' of the wheel and divided into ten hues labelled B, BG, G, GY, Y, YR, R, RP, P and PB. Each of the ten hues are themselves divided into ten increments. Saturation increases by equal increments whilst moving radially outwards. The lightness scale range is on the vertical axis and ranges from black at the bottom to white at the top with nine equal intervals (shades of grey) in between.

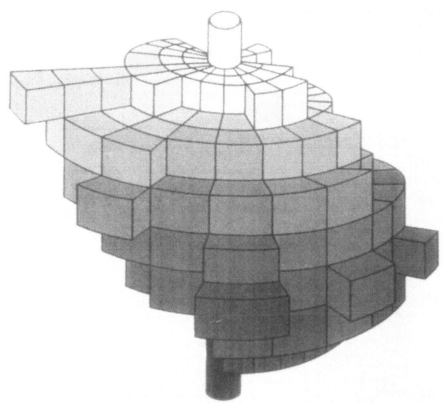

Figure 6.1 Munsell colour solid. Reproduced courtesy of Minolta (UK) Limited. See colour plate section.

6.2 CIE chromaticity diagram

Figure 6.2, reproduced by kind permission of Minolta (UK) Limited[2], shows the CIE chromaticity diagram (CIE = Commision Internationale de l'Eclairage) alternatively referred to as the x–y chromaticity diagram. The axes of the diagram are labelled 'x' and 'y', as with conventional mathematical graphical representations. Lamp specifications include 'x' and 'y' co-ordinate values which enable the user to determine the colour output of the lamp when it has reached luminous stability. Changes in hue and saturation are also shown on the diagram.

6.3 Planckian radiator

It is well documented that most bodies when subject to heat will emit light, initially red but turning more white as the heat is increased. One

Figure 6.2 CIE chromaticity diagram. Values on perimeter of diagram are wavelengths of spectral colours (in nanometres). Point A represents the colour denoted by the co-ordinates x = 0.49, y = 0.30. Reproduced courtesy of Minolta (UK) Limited. See colour plate section.

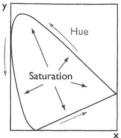

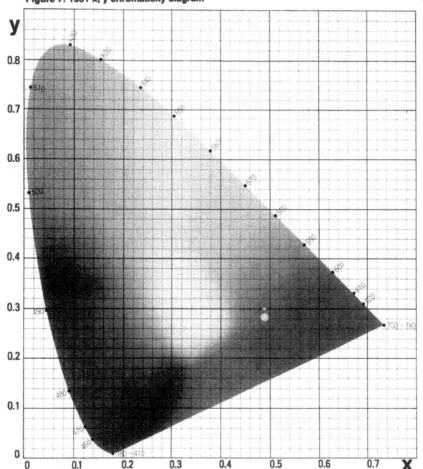

Figure 7: 1931 x, y chromaticity diagram

radiation principle due to Planck shows the calculation of power radiated throughout the electromagnetic spectrum by a full radiator, given in terms of the absolute temperature (Kelvin).

The spectral power distribution (SPD) curves for full radiators at various temperatures are as shown in Figure 2.3, from which it will be

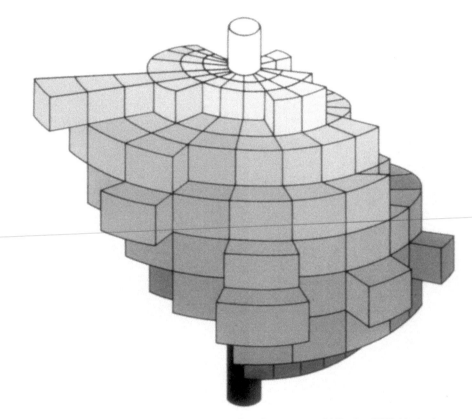

Figure 6.1 Munsell colour solid. Reproduced courtesy of Minolta (UK) Limited.

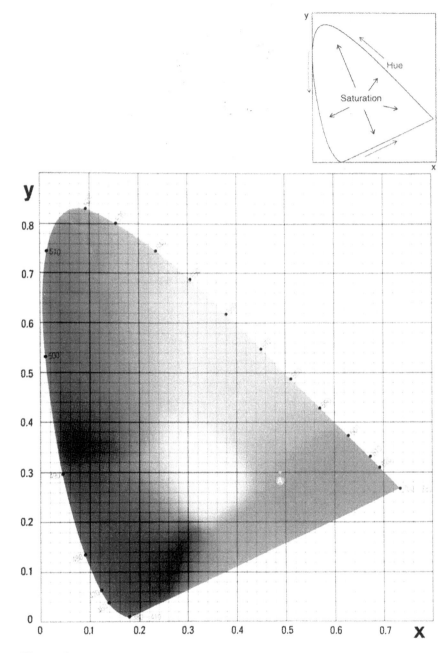

Figure 6.2 CIE chromaticity diagram. Values on perimeter of diagram are wavelengths of spectral colours (in nanometres). Point A represents the colour denoted by the co-ordinates x = 0.49, y = 0.30. Reproduced courtesy of Minolta (UK) Limited.

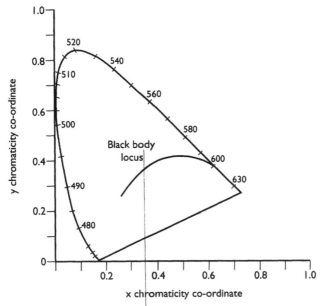

Figure 6.3 Outline of CIE chromaticity diagram showing black body locus. Values on perimeter of diagram are wavelengths of spectral colours (in nanometres).

apparent that the maximum power radiated increases markedly with temperature whilst simultaneously the wavelength at which the maximum power occurs becomes progressively shorter.

If the colour co-ordinates from the black body spectral distribution curves are plotted on a chromaticity diagram, it is found that they all lie on a smooth curve, this being referred to as the *full radiator locus*. Figure 6.3 shows the black body locus superimposed on to the outline of the CIE chromaticity diagram.

If a particular light source of interest has a chromaticity lying directly on the full radiator locus it is said to have the same colour temperature as that particular radiator, even though the spectral power distributions of the two sources may differ markedly. Any light source of interest which does not lie directly on the full radiator locus can be described in terms of its *correlated colour temperature*. Section 6.7 explains in detail the essential differences between colour temperature and correlated colour temperature.

6.4 Dominant wavelength

In addition to the specifying of colour using 'x' and 'y' co-ordinates, it is also possible to refer to the dominant wavelength of colour using the CIE chromaticity diagram, see Figure 6.2. Saturated colours are shown

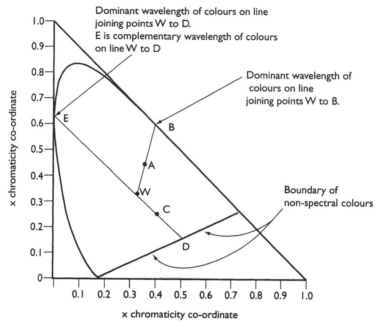

Figure 6.4 Dominant wavelength.

at the perimeter of the diagram, i.e. at the spectrum wavelengths. As the colour moves inwards from the perimeter towards the centre of the diagram, the colour becomes progressively less saturated.

Consider Figure 6.4. A line is drawn joining point W, representing the colour white, to the perimeter of the diagram at point B. The point of intersection of this line with the perimeter, point B, is known as the dominant wavelength of that hue. All of the colours that lie on this line, for example at point A, are assumed to have the same hue. Lines from point W to different values of dominant wavelength will have correspondingly different hues.

Further consider a hue represented by the line joining point W to point D. For any colour on the purple side of point W, for example point C, the dominant wavelength will be represented by point E which is the line joining points W and C produced in the 'opposite' direction. This process is necessary since point D lies on the boundary on non-spectral colours and so therefore it is impossible to have a dominant wavelength on this boundary. It is therefore necessary to use the complimentary wavelength, i.e. point E.

6.5 Chromaticity of the visible spectrum

The spectral colours are identified by their corresponding wavelengths which are shown on the perimeter of the locus shown in Figure 6.2.

6.6 Lamp colour appearance

Colour appearance is the term given to the colour that an object or a light source appears to be. Electric lamps can be classified, inter alia, according to their colour appearance group, details of which are given in Table 6.1.

Table 6.1 Lamp colour appearance groups

Colour temperature or correlated colour temperature range (Kelvin)	Description	Examples
Up to 3300	Warm	S, I, HS FD (some), XF (some)
Above 3300 but not greater than 4000	Intermediate	QE, FD (some) XF
Above 4000 but not greater than 5300	Intermediate-cool	M, FD (some)
Above 5300	Cold	Artificial daylight (FD) Northlight (FD)

N.B. For details of lamp abbreviations used please refer to Table 7.2.

6.7 Colour temperature and correlated colour temperature

The colour temperature of a light source is described as the absolute temperature of a full radiator, or perfect black body, which emits radiation of the same chromaticity as the body under investigation. Consider the simple tungsten filament lamp. The colour output of the lamp approximates to the chromaticity of a black body at the same temperature of the filament itself. The colour temperature of the lamp therefore approximates to the actual temperature of the filament.

When considering discharge lamps it is usual to refer to the correlated colour temperature (CCT). This is the term given to the temperature of a black body having a chromaticity closest to that of the light source under investigation. Consider the example where the colour of a discharge lamp approximates to that of a black body at a temperature of say 3600 Kelvin. The term colour temperature is inappropriate and the lamp is said to have a correlated colour temperature (CCT) of 3600 Kelvin.

It is possible to estimate the value of correlated colour temperature using the CIE chromaticity diagram. Figure 6.5 shows that part of the

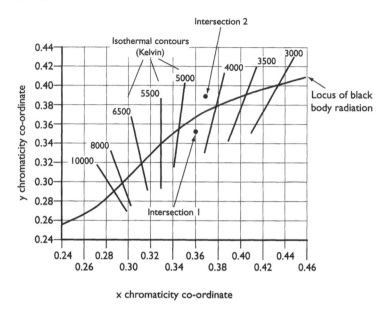

Figure 6.5 Isothermal contours on CIE chromaticity diagram.

diagram in the region of the black body locus. The diagram shows isothermal contours which pass through the locus of black body radiation. The values of x and y chromaticity co-ordinates of a light source under investigation are plotted on the diagram. If the intersection of the co-ordinates falls directly on one of the isothermal contours plotted, then the corresponding value of CCT is that shown on the contour. If, however, the point of intersection of the co-ordinates of the light source under investigation does not lie directly on an isothermal contour, then the corresponding value of CCT is found by interpolation.

Consider discharge lamp source 'A' which has 'x' and 'y' chromaticity co-ordinates of 0.360 and 0.350 respectively, as shown by intersection 1 on Figure 6.5. The intersection does not lie on one of the given isothermal contours. By the process of interpolation the value of CCT is estimated at 4500 Kelvin.

However when specifying light source colours using CCT values, it is possible to have two different sources which exhibit identical values of CCT but show markedly different values of 'x' and 'y' co-ordinates.

For example light source B also has a CCT value of 4500 Kelvin but the values of the 'x' and 'y' chromaticity co-ordinates, i.e. x = 0.370 and y = 0.390, are quite different to those from source 'A', as plotted at intersection 2 on Figure 6.5. Specifying a light source using CCT only can therefore be very misleading.

Table 6.2 gives typical values of colour temperature or correlated colour temperature of various light sources.

Table 6.2 Typical values of colour temperature or correlated colour temperature of various light sources

Lamp description	Typical values of colour temperature ● or correlated colour temperature ◆ (in Kelvin)
Tungsten filament	2800 ●
Tungsten halogen	3000 ●
Fluorescent	Various ◆
H.P. Mercury	3300 to 3800 ◆
Mercury blended (with filament)	3500 ◆
Metal halide	Various ◆
L.P. sodium	Monochromatic
H.P. sodium	2000 to 3000 ◆
Induction	3000 to 4000 ◆

6.8 Colour rendering and colour rendering index (CRI)

It is important to consider the effects on the surface colour of an object when seen under two different light sources. If the two different light sources have different spectral power distributions then the resulting spectral power distribution of the light reflected from the surface of the object under inspection will appear markedly different.

In general terms the product of the spectral power distribution of the illuminant and the spectral reflectance of the surface of the test object will give the spectral power distribution of the test object. Consider for example a post box which is generally accepted as being 'red' in colour. If the post box is illuminated first by a Standard Illuminant (D_{65}), and second by a Standard Illuminant A, then the corresponding spectral power distribution (SPD) of the reflected light, and that which is detected by the eye, will be different.

Consider Figures 6.6, 6.7 and 6.8. Figure 6.6 represents the variation in spectral output of a light source, and Figure 6.7 represents the spectral reflectance properties of an object which is illuminated solely by the source whose spectral power distribution is shown in Figure 6.6. If individual values of SPD of the illuminant and spectral reflectance of the object are multiplied at each value of wavelength, then the spectral power distribution of reflected light from the object will be as shown in Figure 6.8.

Lamp types are categorized according to their colour rendering properties and CRI values all as shown in Table 6.3. Colour rendering index (CRI) values are numerical values which are assigned to the degree of colour rendering provided by a light source. The numerical scale used ranges from zero, which represents a light source where colour is totally distorted, to 100 which represents no colour distortion.

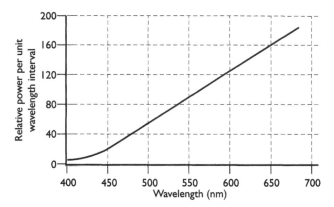

Figure 6.6 SPD of light source used.

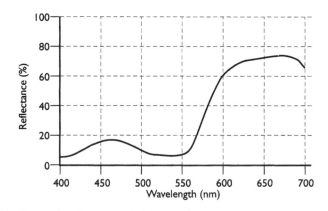

Figure 6.7 Spectral reflectance of object under test.

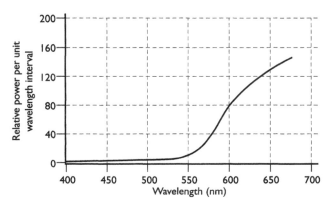

Figure 6.8 SPD of reflected light from object under test.

Table 6.3 Colour rendering properties of lamps and CRI values

Colour rendering group	CRI value	Description	Examples
	100		
IA		Excellent colour rendering	I, HS, northlight (FD) Artificial daylight (FD)
	90		
IB		Very good colour rendering	Some FD, XF
	80		
2		Good colour rendering	Some FD, M Some S
	70		
	60		
3		Poor colour rendering	QE, some FD
	50		
4		Very poor colour rendering	Some S
	40		
	30		
	20		
	10		
			LS
	0		

6.9 Colour contrast

It will be evident that colour contrasts will be influenced by the colour properties of the light source under which surfaces are viewed. If the colour output of the source is (or approximates to) white light, then the subsequent colour rendering and light diffusion will be improved. Table 6.4 shows colour contrasts in descending order of rank.

Table 6.4 Colour contrasts in descending order of rank

Ranking number	Object colour	Background colour
I	Black	Yellow
2	Green	White
3	Red	White
4	Blue	White
5	White	Blue
6	Black	White
7	Yellow	Black
8	White	Red
9	White	Green
I0	White	Black

Source: ILO Encyclopaedia of Occupational Health and Safety, 1998 4th ed. Vol II.[3]

6.10 Colour perception and production

When performing visual tasks during normal daylight, or under suitable and sufficient artificial lighting, an individual will see mainly by using a very small zone of the retina termed the fovea. The cone-shaped light-sensitive nerves in this region have a response which differs according to the spectral colours.

When an observer views a red surface, the red-sensitive cones will be stimulated and they subsequently convert this sensation into electrical signals which are then conveyed to the brain. A similar situation occurs when the observer views either a blue or green surface. In practice, the vast majority of colours are not confined to monochromatic wavelengths but consist of a combination of the colours of the spectrum. The individual colour-sensitive cones will detect their own surface colour and the corresponding signals are combined in the brain so as to convey the impression of the true surface colour of the object under investigation.

Several mechanisms are present which cause the human eye to detect that which we have come to refer to as colour. The combination of the eye and the brain may allow us to detect a particular sensation which we refer to as a colour but only by using instrumentation (such as a spectrometer) can an accurate record be made of the complete spectral composition.

When the luminous intensity is low, as typified by a dark night, the human eye fails to perceive colour. The inaccurate physiological response of the human eye to the spectral information provided by an object is typified by the experiment involving Benham's disk, as shown in Figure 6.9. If the disk is rotated at an angular velocity of say two revolutions per second in daylight, the black and white areas will apparently disappear and the eye perceives colours on the disk, the outer arc

Figure 6.9 Benham's disk. Rotation of disk at 2 rps produces a sensation of colour in the arcs.

seemingly yellow and the inner arcs blue-green and black. This phenomenon is caused by, inter alia, deficiencies within the human eye.

Despite these phenomena, resulting in a failure of perception, there are in the main four mechanisms by which colours are produced:

- colours produced by *emission*, e.g. firefly and colour television;
- colours produced by *interference and diffraction*, e.g. soap bubbles and oil film on a pool of water;
- colours produced by *scattering and dispersion*, e.g. the sky and the sunset; and
- colours produced by *selective absorption of wavelengths*, e.g. an orange and an apple.

6.11 Colour mixing and complementary colours

The effect produced when colours are mixed depends upon whether the mixture involves lights or paints. Figure 6.10 shows the effects produced. Colour mixing involving lights is termed *additive* and that involving paints is referred to as *subtractive*.

Thus, if say a blue light from an artificial light source is directed onto a white screen and then green light from a second artificial light source is superimposed onto the same screen, the resultant colour is cyan. Similarly a mix of red and blue lights from artificial light sources will produce a magenta colour.

By using a suitable mix of the three colours listed, i.e. red, green and blue together with white, any desired colour can be produced. The three colours red, green and blue are referred to as *primary* colours.

If two colours of light are mixed (additive mixing), the resultant colour will lie on the straight line joining the chromaticity co-ordinates of the component light sources. Reference to Figure 6.10 shows that white light can be made up of mixtures of different spectral wavelengths. Thus, for example, white light can be made up of mixtures, in correct proportions, of green and magenta, red and cyan or blue and yellow.

Two spectrum colours which, when combined, form white light are said to be complementary to each other and a *complementary* colour can be defined as that which, when combined with another, will produce white light. The complement of a primary colour is that secondary colour, i.e. non-primary colour, which is produced as a result of combining the other two primaries. The complement of a secondary colour is therefore that primary colour which it does not contain.

In respect of lights, the complement of red is the product of combining blue and green which is cyan. Similarly the complement of blue is the product of combining red and green, i.e. yellow, and the complement of green is the product of combining red and blue which is magenta.

The primary colours in respect of paints are the complements of the primary colours of light.

Additive colour mixing (lights)

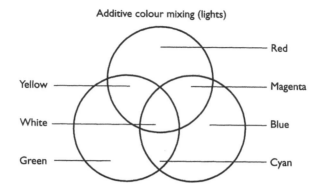

Subtractive colour mixing (paints)

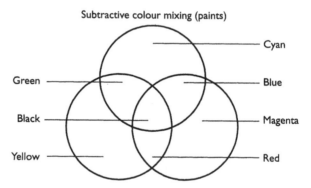

Figure 6.10 Additive and subtractive colour mixing.

6.12 Colour coding

The surface colours of objects can play a significant role in the process of identification. Distinctive colours or carefully selected colour contrasts can be used to advantage in order to alert operatives to pay some form of special attention.

Two types of colour coding are employed and they are termed *connotative* and *denotative*.

Connotative colour coding refers to the situation where object colours convey a message when no other characteristics are present on the object under consideration which would aid identification. An example of connotative colour coding is the identification of electrical resistors. Denotative colour coding refers to the situation where the information conveyed is not restricted to the use of surface colours but can also be found by other means. Advantages of the use of denotative colour coding include improved response times and a reduction in errors made in obtaining information. An example of denotative colour coding is identification of different clothes sizes in a store by virtue of different

Table 6.5 Colour coding for industry

- **Atlases and maps**
  Road atlases and maps use colours to differentiate between roads etc.
  The map of the London Underground uses colours to denote different lines.
- **Electrical components**
  Electrical resistors are colour coded so that identification of the value of the
  resistor, in ohms, can be recognized without the necessity for using measuring
  equipment.
  Table 6.6 shows the colour coding system used for electrical resistors.
- **Fire and safety equipment**
  Coloured graphic displays are also used on walls adjacent to the equipment.
- **Machinery**
  Stop and Emergency Stop buttons should be made conspicuous by the use of
  bright colours.
- **Medical gases**
  Anaesthetic gases used in hospitals, dental surgeries etc., are colour coded.
- **Pipework and tubing**
  Colour coding is extremely beneficial when pipes and tubes are used to transport
  dangerous substances.
- **Staircases and stairways**
  Markings are essential and must be in accordance with relevant legislation.

colours of coat hangers used. Other examples of situations where colour coding is used are given in Table 6.5.

Thus, for example, if a resistor is marked as shown in Figure 6.11, then using the colour code shown in Table 6.6, the sequential combination of green, white, red and silver will have a resistance value (in ohms) of:

green = 5 (first digit)
white = 9 (second digit)
red   = 2 (multiplier i.e. number of zeros)
silver = 10 per cent (tolerance).

The total resistance value = 5900 ohms (or 5.9 kΩ) ±10 per cent.

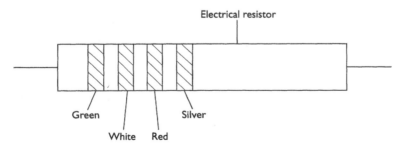

Figure 6.11 Electrical resistor of value 5900 ohms ±10 per cent.

Table 6.6 Colour coding of electrical resistors

Colour of band	Value
Black	0
Brown	1
Red	2
Orange	3
Yellow	4
Green	5
Blue	6
Violet	7
Grey	8
White	9
Gold	±5%
Silver	±10%
No colour	±20%

6.12.1 Colours used for transport signals

Coloured signals are used in road, rail, marine and aviation transport signalling, typically involving red, yellow, amber, green, blue and white colours. The colour of transport signals can be specified by defined areas on the CIE chromaticity diagram.

Ships are required to show coloured lights, red on the port side and green on the starboard side, which will allow other marine traffic in the vicinity to be aware of the direction of travel. Similarly colour signals lamps are used on aircraft, red on the port wing and green on the starboard wing. Flashing red lights are used as anti-collision warning signals. Runway lighting also involves coloured signal lamps and beacons to convey information to pilots.

For rail transport, the direction of travel of trains is indicated by coloured signal lamps, white on the front and red on the rear. Traffic signalling on railways, as with road traffic, uses red, amber and green signal aspects.

6.13 Metamerism

It is well documented that the surface colour of an object depends upon the characteristics of the light source under which the surface is viewed. One problem which is related to this phenomenon is that of viewing two objects whose surface colours appear the same under say daylight but yet appear very different under certain forms of artificial lighting. This phenomenon is referred to as metamerism. With metameric objects, the spectral reflectance characteristics of the two colours of the objects are different whereas the corresponding tristimulus values (representing contributions from red, green and blue) will be the same under one light source but different from each other under another light source. This usually occurs as a consequence of the use of different pigments or materials.

Workers involved in the inspection of finished products where colour discrimination is important should be aware of the potentially misleading effects produced by metamerism.

6.14 Photochromism

Chemical compounds that undergo reversible colour changes when they are exposed to light are said to exhibit photochromism. Consider Figure 6.12 where curve 1 represents the absorption spectrum of material X. When material X is irradiated in this region of the spectrum, typically the wavelength of irradiation (λx) is in the ultraviolet region, the material is converted into material Y which has an absorption spectrum as shown by curve 2 which is in the visible region of the spectrum. The reverse process may be achieved by heat or alternatively by irradiation of material Y with light in the wavelength region of λy. Photochromism can be organic or inorganic and inorganic photochromism, based on glasses which contain silver halide, is used in the manufacture of photochromic spectacles. Such spectacles have lenses which become darker in bright environments and less dark in dimmer environments.

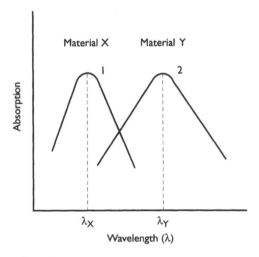

Figure 6.12 Photochromism.

6.15 Achromatic interval

If coloured lights are gradually diminished in intensity they appear colourless for a small range of intensity values prior to being extinguished. This range is referred to as the *achromatic interval*. The effect can be shown under laboratory conditions using a simple incandescent lamp supplied from a variable voltage supply. The lamp should be suitably covered so as to produce a diffuse light. If the lamp has been operating on rated voltage and the voltage is then reduced, the light from the lamp will appear red, gradually becoming more faint, and just before the lamp extinguishes it appears grey.

6.16 Effects of environmental colour on individuals

The colour of light, its corresponding distribution and the colours subsequently used in a particular environment are, inter alia, significant factors which will influence the sensations felt by individuals. Some of the principles upon which colour scheme design is often based include:

• Bright colours will usually tend to produce an aesthetically pleasing environment which is stimulating whilst dark colour schemes have the opposite effect which can lead to depression.
• Warm colours can tend to excite the nervous system and can lead to the feeling that the temperature is rising.
• Cold colours can have the opposite effect in that they can have a calming influence.
• The sensation of colour of an object is influenced by the background colour and also by the effect that a light source will have on its surface.
• By using either warm or cold lighting, the effects of hot or cold physical environments can be compensated.

Chapter 7

Natural and artificial light sources

7.1 Daylight

Fundamentally daylight is derived from the sun. The temperature of the sun at its surface is accepted as lying in the range 6000 Kelvin to 7000 Kelvin.

The sun's surface approximates to that of a black body or full radiator. Losses by scattering produce a substantial movement away from full radiation. When compared to a black body (or full radiator) daylight is found to be deficient of wavelengths typically less than 430 nm.

When the sky is cloudy the height of the clouds will affect the resultant spectral distribution of daylight. When the cloud cover is relatively high, the spectral power distribution perceived is similar to that of the sun outside the earth's atmosphere. The resulting colour temperature which approximates to that of the spectral power distribution applying on cloudy days is typically in the region of 6500 Kelvin.

7.2 Availability of daylight

In the United Kingdom, the prevailing external illuminance due to daylight reaches a maximum of approximately 35 000 lux at noon during the month of July. During December the peak value at noon is approximately 8000 lux. When the prevailing external illuminance value falls below 5000 lux, daylight is usually assumed to be insufficient to sustain suitable internal lighting.

It will be evident that the illuminance due to daylight within a room interior is subject to the same pro-rata variations that occur externally due to daylight. The prevailing illuminance at any given point within an interior can be shown to be, to a first approximation, a constant fraction of the simultaneous prevailing illuminance due to daylight at a fixed point external to the building. This fraction is known as the *daylight factor* and it is usually expressed as a percentage value. Daylight factor is considered in detail in Section 7.3.

There will be a finite time during each day when the level of daylight becomes insufficient for occupants of buildings to perform visual tasks. At this time artificial lighting has to play the major role in providing suitable and sufficient lighting within an interior. It will be further evident that the time at which artificial lighting must be provided in interiors is influenced by the amount of penetration of natural daylight into interiors.

7.3 Daylight factor

The level of illuminance within an interior, due to natural daylight, varies in accordance with the changes in prevailing external illuminance due to daylight.

The parameter daylight factor (DF) is calculated from the relationship:

$$\text{Daylight factor (DF)} = \frac{\text{Illuminance due to daylight at a particular point within a room}}{\text{Simultaneous illuminance on a horizontal plane external to the room from a completely unobstructed and overcast sky}} \times 100\% \qquad (7.1)$$

It is also possible to represent daylight factor by the expression:

$$\text{Daylight factor} = \frac{\text{Sky component}}{(\text{SC})} + \frac{\text{Externally reflected component}}{(\text{ERC})} + \frac{\text{Internally reflected component}}{(\text{IRC})} \qquad (7.2)$$

The concepts of the three components referred to, i.e. sky component, externally reflected component and internally reflected component, are shown in Figure 7.1. It will be clear that the value of daylight factor at a particular point in a room is constant. Instrumentation is available which allows daylight factor to be measured directly.

Daylight factor values are individual values relative to individual points within an interior and their use as such may be considered limited. It is far more meaningful to refer to the *average daylight factor* within an interior. The value of average daylight factor can be found from the relationship:

$$\text{Average daylight factor} = \frac{0.85W\theta}{A\,(1 - R^2)} \qquad (7.3)$$

where:

W = total glazed area of window (m²);
A = total area of all surfaces (m²);
R = average reflectance value of all room surfaces;
θ = angle (in degrees) in the vertical plane subtended by the sky visible from the centre of a window.

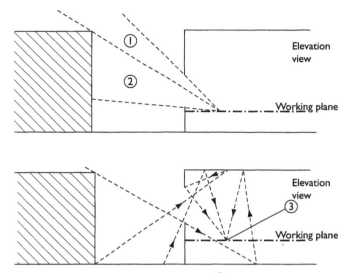

Figure 7.1 Three components of daylight factor. ① Sky component (SC); ② Externally-reflected component (ERC); ③ Internally-reflected component (IRC).

7.4 Emission of light from heated bodies

If heat is applied to a metal it will be observed that visible light is emitted provided that certain conditions are satisfied. The light emitted is initially a dull red in colour and if sufficient heat is applied the material appears white. An everyday example of the use of the emission of light from a heated body is the tungsten filament lamp (GLS).

When the temperature of a body is relatively low it emits only long wavelength radiation. When the temperature increases, light of progressively shorter wavelengths will be emitted until all of the wavelengths in the visible range are included.

Heat radiated from a body depends, inter alia, on the characteristics of the body surface. A dull black body will radiate better, at a given temperature, than a shiny body. It is often advantageous to consider the radiation from a perfect black body to which heat is applied. The term perfect *black body radiator*, also known as a *full radiator*, is considered in Section 2.5.

In reality there are no such objects as black bodies, although the radiation from some heated bodies, for example a tungsten filament, approximates to black body radiation.

Figure 7.2 shows the variation in relative radiated energy emitted by certain materials used for lamp filaments all at the same temperature and plotted against a base of wavelength.

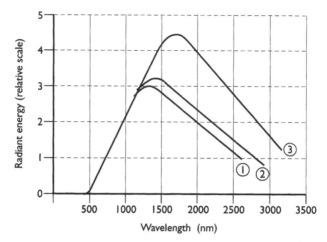

Figure 7.2 Relative radiated energy for various materials used as filaments – all materials are at the same temperature. ① Osmium; ② Tungsten; ③ Carbon.

7.5 Types of spectra

The light emitted from a very hot solid body is often termed 'white light' since it contains all of the colours of the visible spectrum. The light output from a tungsten filament lamp similarly encompasses a band of wavelengths ranging from red through to violet, although the SPD of a tungsten filament lamp shows a bias towards the red end of the visible spectrum. Such a spectrum is termed a *continuous spectrum*.

When an electrical discharge occurs in a gas or vapour, the spectrum so produced is discontinuous. Spectral analysis of the output produced shows that sharp spectral lines are emitted at individual wavelengths. Different gases and vapours produce different spectra which vary both in the number of lines of individual wavelengths and subsequently the wavelength and hence the colour of the lines. Spectra which are discontinuous and contain a number of individual wavelength contributions are referred to as *line spectra*.

In a neon lamp the output contains a sequence of lines in the red region of the visible spectrum which accounts for the red output from the lamp. Additionally in a sodium lamp there is a yellow line. Any light emitted by sodium, or any of the chemical compounds containing sodium, will always have a characteristic yellow tint in their output. Heating a few grains of common salt in a gas flame will produce the desired effect.

The lines emitted by various materials are often referred to by letters of the alphabet, for example the output from the low pressure sodium discharge is referred to as the D line. Lines referred to in this manner are known as *Fraunhofer lines*. The principal lines are shown in Table 7.1.

Table 7.1 Principal Fraunhofer lines

Line designation	Element to which line corresponds	Perceived colour	Wavelength (nm)
A	O	Deep red	759.4
B	O	Red	686.7
C	H	Red	656.3
d	He	Yellow	587.6
D_1	Na	Yellow	589.6
D_2	Na	Yellow	589.0
E	Fe & Ca	Green	527.0
b	Mg	Blue–Green	518.4
F	H	Blue	486.1
G	H	Blue	434.0
G	Fe & Ca	Violet	430.8
H	Ca	Violet	396.9
K	Ca	Violet	393.4

7.6 Emission of light from gas discharges

With discharge lamps the light output is produced as a result of an electrical discharge. Such a lamp of necessity consists of a discharge tube which contains a gas or vapour. When a voltage is applied to electrodes, located in the ends of the discharge tube, energy is transferred to the atoms of the gas within the tube. Electrons which are orbiting the nuclei of the atoms are then displaced to a higher energy level orbit. They remain in this higher energy level orbit for typically less than one millionth of a second, after which they return to the lower energy level orbit from which they had originally been raised. When the electrons fall back to the lower energy level orbit they simultaneously emit the excess energy, which they had gained in being raised to the higher energy level orbit, in the form of photons of light, as shown in Figure 7.3. The properties of the gas or vapour contained within the discharge tube will influence the wavelengths of the light produced and hence the colour output of the lamp. Such a lamp will emit line spectra.

Discharge lamps will, of necessity, require electrical and/or electronic control equipment which serves two essential functions. The equipment must be capable of developing and supplying a relatively high voltage in order to assist the starting of the lamp and then once the arc has been struck and becomes established, the equipment takes on the function of a current limiting device in the electrical circuit. It is convenient to sub-divide discharge lamps into low and high pressure categories.

Unfortunately discharge lamps do not attain their maximum light output immediately following switch on from a cold start. This includes fluorescent lamps, although with such lamps the fraction of maximum possible light output, available immediately following switch on, is usually relatively high.[4]

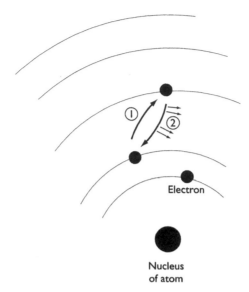

Figure 7.3 Emission of light from gas discharges. ① Application of energy displaces electron to higher energy level orbit; ② Electron remains in higher energy level orbit for typically nanoseconds before returning to its original energy level orbit. In doing so, light is emitted, the wavelength of which depends upon the difference in energy between the two orbits.

7.7 Tungsten lamps

This group of lamps, often referred to as *incandescent lamps*, include tungsten filament and tungsten halogen lamps.

7.7.1 Tungsten filament lamps

The general lighting service (GLS) lamp, as shown in Figure 7.4, is a simple tungsten filament lamp. The lamp filament, when heated to a temperature of approximately 2800 Kelvin, becomes white hot and emits radiation throughout the visible spectrum with a bias towards the higher wavelengths. The outer glass envelope is filled typically with a mixture of nitrogen and argon whose function is to limit the evaporation of tungsten from the filament and, additionally, to assist in the prevention of arcing across the lamp filament. The outer glass envelope may be either clear or alternatively having a pearl finish. In clear lamps the filament is visible and there is normally an abrupt cut-off of light at certain angles. Pearl lamps are often preferred when the lamp is visible in use. The only difference between the pearl lamp and the clear lamp is that the pearl lamp envelope is subjected to internal acid-spraying processes which roughen the inner surface of the glass and in so doing form a multitude of minute prisms, each of which refracts the light in a

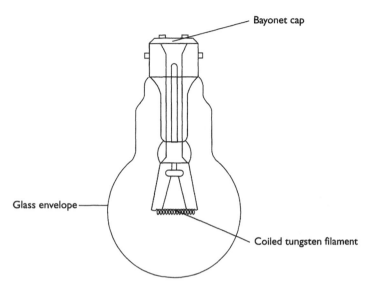

Figure 7.4 General lighting service (GLS) lamp.

different direction. The overall effect is to create diffusion which conceals the lamp filament.

Daylight simulation lamps have been introduced which are typically tungsten filament lamps with a blue glass envelope. Theoretically the combination of the light emitted from the tungsten filament and the effects of the blue glass envelope give an overall output which simulates daylight.

The GLS lamp is dimmable, and it will operate in any position. The lamp produces an instantaneous response to being switched on and further it has instantaneous re-strike capabilities. The tungsten filament lamp can be connected directly to the mains supply without the necessity for electrical and/or electronic control equipment.

Smaller versions of tungsten filament lamps often use krypton instead of argon for the gas filling. Krypton has a lower specific heat and a higher density than argon, which allows higher filament temperatures to be achieved with less heat loss to the gas with a consequent increase in luminous efficiency.

7.7.2 Tungsten halogen lamps

The outer envelope of a tungsten halogen lamp is made of quartz. The filament can operate safely at a much higher temperature than that achieved with a GLS lamp, typically 3000 to 3400 Kelvin. Furthermore the pressure of the gas within the quartz outer envelope can be increased from that applying in the GLS lamp.

During production of the lamp, a small quantity of one of the halogen group of elements (typically iodine, bromine or chlorine) is added to the

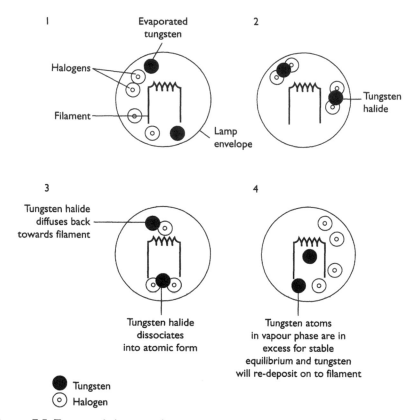

Figure 7.5 Tungsten halogen cycle.

gas filling. Under normal working conditions the tungsten halogen cycle will occur. The halogen combines with the evaporated tungsten, which has become detached from the filament, to form a tungsten halogen. As a consequence the tungsten will not be deposited on the inside of the quartz outer envelope. When the circulating tungsten halogen comes into contact with the filament, which is at a much higher temperature, the tungsten is re-deposited onto the filament and the halogen is then released. One essential requirement of operation of the halogen cycle in the tungsten halogen lamp is that the lamp wall temperature must be maintained at a minimum value of 250°C for optimum operating conditions.

Figure 7.5 shows the principle of the tungsten halogen cycle and Figure 7.6 shows the constructional features of typical tungsten halogen lamps.

The tungsten halogen lamp is dimmable, although this may cause problems with maintaining the lamp wall temperature at 250°C. The lamp will operate in various positions and it produces an instantaneous response to being switched on and further it has instantaneous re-strike capabilities. The tungsten halogen lamp can be connected directly to the

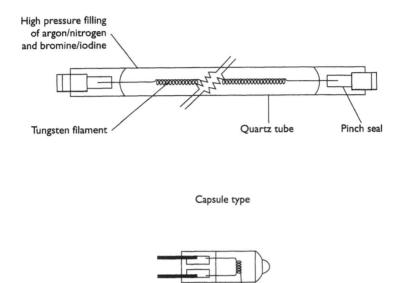

Linear type (double-ended)

High pressure filling
of argon/nitrogen
and bromine/iodine

Tungsten filament   Quartz tube   Pinch seal

Capsule type

Tungsten filament

Figure 7.6 Tungsten halogen lamps.

mains supply without the necessity for electrical and/or electronic control equipment.

When compared with the GLS lamp the tungsten halogen lamp emits greater UV radiation, due to the increase in operating temperature coupled with the increase in transmission of the ultraviolet through the quartz envelope. The quartz envelope of a tungsten halogen lamp should not be touched by bare hands. The acids and greases which are present on the fingers will attack the quartz. This ultimately forms weak spots which lead to subsequent premature failure of the lamp.

The tungsten halogen lamp operates with an internal pressure above atmospheric and it is possible for the lamp to shatter. Due care must therefore be taken to ensure that lamp fragments will not cause damage to individuals. Luminaires should be so designed that fragments should be contained.

7.7.3 Dichroic reflector lamps

One of the major problems which accompanies the use of tungsten halogen lamps is that of excessive heat emitted directly in front of the lamp. This will have adverse effects on any material objects in close proximity to the front of the lamp. In display advertising this can be potentially damaging since the heat emitted by the lamp is likely to create a fire hazard.

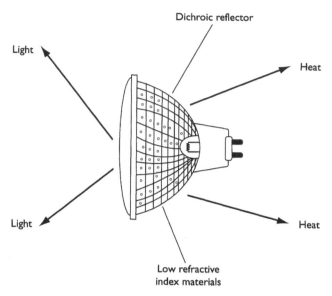

Figure 7.7 Dichroic reflector lamp.

Low voltage dichroic reflector lamps have a been introduced. These have a special type of reflector which transmits outwards (from the front of the lamp) mostly visible radiation, whilst simultaneously allowing the infrared radiation (emitted as part of the light producing process) to pass through the reflector and out through the back of the lamp. The constructional details of the dichroic reflector are as shown in Figure 7.7.

The basic principle of operation of the dichroic filter relies upon the production of 180° phase shifts of incident wavelength at the boundary layers since, as the light produced passes through each layer, interference effects will occur. It follows that certain incident wavelengths will be reflected whilst others will be transmitted.

7.8 Discharge lamps

Incandescent lamps rely for their operation on the passage of an electrical current through a metallic filament which becomes heated and emits radiation. Discharge lamps, however, rely on the passage of an electrical current through a gas or vapour for their operation.

7.8.1 Introduction

The general principle of operation of all discharge lamps, with the exception of the induction lamp, is explained in detail in Section 7.6. Figure

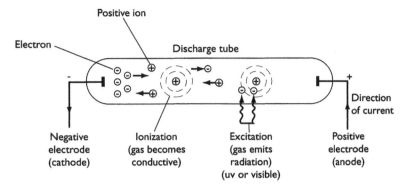

Figure 7.8 Light production in discharge lamp.

7.3 shows the emission of light from gas discharges and Figure 7.8 shows the application of this principle to practical discharge lamps, which rely on ionization and excitation for their operation.

The types of discharge lamps currently available include the following:

- low pressure mercury or fluorescent;
- high pressure mercury;
- metal halide;
- low pressure sodium;
- high pressure sodium;
- mercury blended; and
- induction.

Figures 7.9, 7.12, 7.13, 7.14, 7.15, 7.16, 7.18 and 7.19 show the constructional features of the lamps considered.

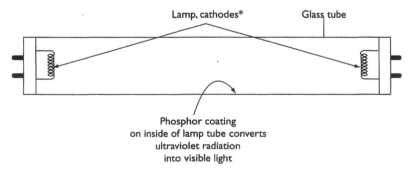

Figure 7.9 Tubular fluorescent lamp. *The term cathodes is often used when referring to both electrodes, even though they take on the roles of anode and cathode alternately.

7.8.2 Low pressure mercury or fluorescent lamps

In the fluorescent lamp, the electrical discharge produced has two major spectral lines. Approximately 85 per cent of the output is at a wavelength of 254 nm and approximately 15 per cent of the output is at a wavelength of 185 nm. Both of these spectral lines are in the UVC band. It is necessary therefore to convert these incident wavelengths into wavelengths in the visible part of the spectrum. This is achieved by internally coating the glass discharge tube with phosphors which have the ability to absorb incident wavelengths in the UV band and subsequently re-radiate them at longer wavelengths which lie within the visible spectrum. The wavelength of the light produced by a fluorescent lamp is influenced by the composition of the phosphors coated on the inside of the lamp glass wall.

Initially halophosphates were used in order to make high efficiency white lamps and, by the 1970s, phosphor development had led to the introduction of narrow waveband phosphors, which separately emitted red, blue and green light (hence triphosphors), which, when combined, produced white light. Multi-phosphors have enabled an even greater selection of lamp colours.

Fluorescent lamps have several constructional forms, the tubular version (see Figure 7.9) being used extensively. Electronic circuitry allows some tubular lamps to be successfully dimmed in a process which involves high frequency operation.

One major disadvantage of the fluorescent lamp is its operation under reduced temperatures. By way of example, consider a fluorescent lamp operating at an ambient temperature of, say, 0°C. Assuming that the

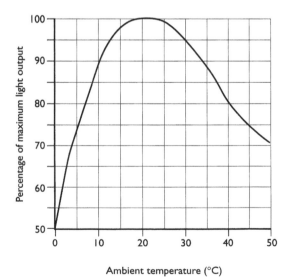

Figure 7.10 Typical effect of ambient temperature on light output of tubular fluorescent lamp.

lamp would strike at that temperature, which is not guaranteed, the light output after the lamp had reached luminous stability, would be typically 50 per cent of that expected if the lamp was operating in a normal room temperature of 20°C, as shown in Figure 7.10. It will be evident therefore that fluorescent lamps are not normally used for external applications or in those internal applications where low ambient temperatures are experienced.

When used in low ambient temperatures, fluorescent lamps will often produce symptoms characterized by regions of high and low brightness of light output from the lamp, appearing as hoops, which may appear to be either stationary, move in either direction along the length of the lamp or even oscillate, as shown in Figure 7.11. This phenomenon is referred to as *striations*, which occur as a consequence of unstable conditions in the gas discharge. They may also appear in new lamps this being due to the mercury not being totally absorbed into the lamp phosphor coating.[5]

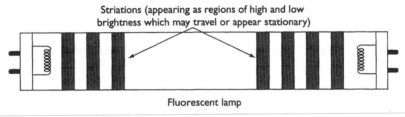

Figure 7.11 Striations in fluorescent lamp.

Special fluorescent lamps which have a predominantly red output have been used for dark adaptation applications. Rods in the retina are relatively insensitive to red light, and, as a consequence, the rhodopsin does not get bleached by light of this colour. The cones can function adequately with red light and it follows that an individual who is indoors and adapted to red light and who then moves swiftly to view a dark night scene externally, will still be able to see adequately as the rods will be able to respond almost immediately. The rhodopsin will still be intact and therefore no period of dark adaptation will be necessary.

Fluorescent lamps will usually re-strike immediately after a momentary loss of electrical supply. Fluorescent lamps can be operated in any position. Figures 7.9 and 7.12 show respectively the constructional features of the tubular and one type of compact fluorescent lamp. Fluorescent lamps operating on high frequency supplies are finding increasing use in interior applications. The major benefits of using fluorescent lamps on such supplies include:

• almost total elimination of flicker, see Section 11.16; and
• the ability to control the light output from the lamp.

If fluorescent lamps, supplied at high frequency, are used in an energy management system, then it is usual to control the light output from

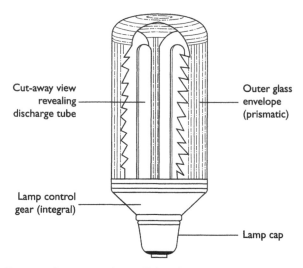

Cut-away view
revealing
discharge tube

Outer glass
envelope
(prismatic)

Lamp control
gear (integral)

Lamp cap

Figure 7.12 Compact fluorescent lamp. Other forms are available.

them automatically in sympathy with the illuminance provided by the prevailing daylight so that in combination the illuminance on the working plane in an interior remains approximately constant. The electrical power consumption of lamps and control gear (operating in combination) is correspondingly reduced when the light output is deliberately dimmed.

Twenty-six millimetre diameter fluorescent lamps use a combination of krypton and argon as the fill gas, typically in a 3 to 1 ratio. The fill gas aids starting in conventional or *hot cathode lamps*, since lamp starting is achieved by pre-heating the electrodes so as to create sufficient ionization of the mercury vapour. *Cold cathode lamps*, often used in advertising signs, rely on a relatively high lamp voltage for establishment of the arc. Electrical isolation of such sign lamps is usually achieved by means of the fireman's switch, which is typically located on shop and store frontages. Lamp life for cold cathode lamps is much greater than for hot cathode lamps.

The electrodes used for cold cathode lamps are typically plain nickel or iron cylinders, whose size is substantial in order to keep the current density at their surface to an acceptably low value.

7.8.3 High pressure mercury vapour lamps

The constructional features of the high pressure mercury vapour lamp are shown in Figure 7.13. At higher pressures the mercury discharge is predominantly concentrated in the blue and green regions of the visible spectrum, together with some ultraviolet wavelengths. The outer envelope of the lamp is usually coated with phosphor, which absorbs the ultraviolet emitted and converts it into wavelengths at the red end of the visible spectrum, where little is emitted by the discharge itself.

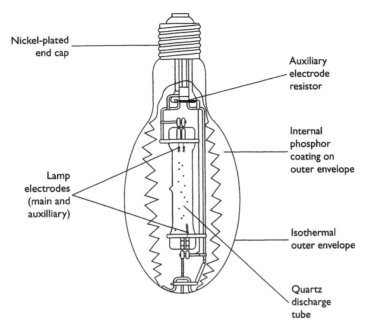

Nickel-plated
end cap

Auxiliary
electrode
resistor

Internal
phosphor
coating on
outer envelope

Lamp
electrodes
(main and
auxilliary)

Isothermal
outer envelope

Quartz
discharge
tube

Figure 7.13 High pressure mercury vapour lamp.

This has the effect of correcting the colour of the lamp and earlier forms of this lamp were often known as 'colour corrected mercury vapour lamps'.

A special type of mercury vapour lamp incorporates a tungsten filament in its construction, as shown in Figure 7.14. The lamp is often referred to as the *mercury blended lamp*. The filament acts as a current limiting device, and in addition, adds some of the 'warm' colours of the spectrum to the output from the mercury discharge. This lamp does not require external control gear for its operation. This lamp takes time to attain its steady-state luminous output, although the inclusion of the tungsten filament means that some useful light output is available immediately following switch on of the electrical supply.

Both the high pressure mercury vapour lamp and the mercury blended lamp are not dimmable and their re-strike capabilities will of necessity involve a time delay. There are few restrictions on the operating positions of such lamps.

7.8.4 Metal halide lamps

If selected materials are added into the discharge tube of a high pressure mercury vapour lamp the colour output can be improved. The dosage of the additives is relatively small and, in order to produce a more accurately-controlled output, it is more appropriate to use the metal in powder form as halides. Some of the halides used include dysprosium,

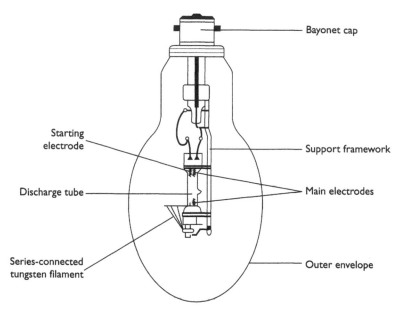

Figure 7.14 Mercury blended lamp.

indium, lithium, scandium, sodium, thallium and tin. Figure 7.15 shows the constructional features of the metal halide lamp.

Metal halide lamps, as with high pressure mercury vapour lamps, are not dimmable. Their re-strike capabilities are poor and involve a time delay and in addition the lamps can only be operated in limited positions.

7.8.5 Low pressure sodium lamps

Figure 7.16 shows the typical constructional features of the low pressure sodium lamp. With low pressure sodium lamps the arc tube, which is made of special ply glass, has an internal coating which is resistant to sodium. Figure 7.16 shows that the arc tube is shaped like the letter 'U' and it is located within an outer vacuum jacket, thereby assisting in the control of thermal stability.

As a consequence of the neon/argon gas filling within the discharge tube, the lamp output is initially a red glow immediately following switch on. As the lamp approaches a steady-state luminous output, the observed radiation becomes monochromatic. Strictly this is a slight misrepresentation since the output consists of two spectral lines at 589 nm and 589.6 nm. It is however usual to consider the output as monochromatic, as shown in Figure 7.17.

The low pressure sodium lamp has a major advantage over other artificial light sources in that it is very useful in foggy or steamy environments. In such conditions, droplets of water, in suspension in the atmosphere,

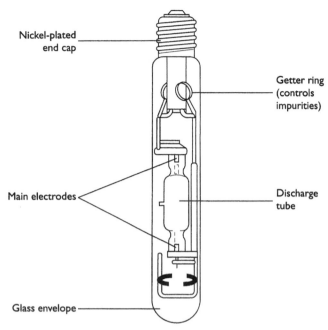

Figure 7.15 Metal halide lamp (tubular).

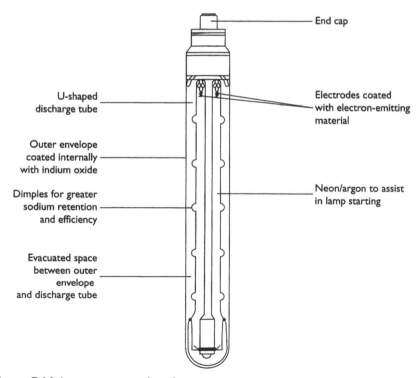

Figure 7.16 Low pressure sodium lamp.

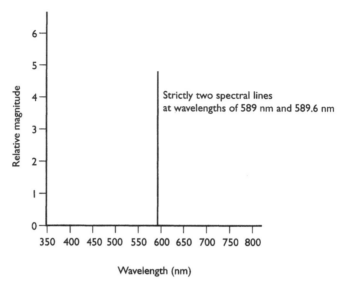

Figure 7.17 Spectral output of low pressure sodium lamp.

act as prisms. The incident light is monochromatic, which will not be dispersed after striking the droplets.

The major disadvantage of the low pressure sodium lamp, and one which renders it unsuitable in many applications, is its colour rendering properties. Surface colours of objects seen in the light output from such a lamp are seriously distorted. The use of the low pressure lamp is therefore restricted to those applications where colour discrimination is not a prime consideration.

The light output from low power ratings of the lamp will not be adversely affected when there is a momentary loss in electrical supply and when the electrical supply is restored the lamp should re-strike almost immediately. For higher power ratings of the low pressure sodium lamp there will be a time delay before the arc will re-strike. Low pressure sodium lamps are not dimmable.

The efficiency of a low pressure sodium lamp diminishes rapidly as the current density is raised above an optimum value. The lamp must therefore be operated at a relatively low current value. It follows that the surface area of the discharge tube must, of necessity, be large for the power dissipated. Low pressure sodium lamps are therefore made with relatively long discharge tubes which are bent into the shape of a letter 'U'. One strange phenomenon associated with such lamps is that excited sodium vapour is opaque to its own radiation, the consequences of which are that light emitted from one limb of the 'U'-shaped discharge tube will not transmit through the other limb. When the lamp is installed horizontally in a luminaire (which is the usual burning position), the lamp should ideally be orientated so that the limbs of the 'U' tube are arranged one above the other, in order to produce the greatest light output from the lamp.

7.8.6 High pressure sodium lamps

If the pressure of the sodium within the discharge tube is raised from that level which applies in a low pressure sodium lamp, the luminous output tends more towards the yellow region of the spectrum. The net effect is that the colour appearance of the lamp shifts towards a golden white. This has the advantage of improving the colour rendering of the lamp, although there is a disadvantage associated with the increase in colour rendering in that there is a reduction in lamp efficacy when compared with the low pressure sodium lamp.

The conventional arc tube cannot be used for high pressure sodium lamps due to the reactive nature of sodium. The discharge tube for a high pressure lamp is therefore constructed from translucent polycrystalline alumina. The outer glass envelope of the lamp is evacuated so as to assist in the prevention of arcing and oxidation.

The lamps are not dimmable and, should the electrical supply to the lamps be momentarily interrupted, a time delay will be involved in restriking. Such lamps can, however, be operated in any position.

Figure 7.18 shows the constructional features of the high pressure sodium lamp.

In the event of a momentary loss of electrical supply to the lamp, the pressure in the discharge tube will have to fall before the lamp arc can be re-established and there will therefore be a further time delay whilst

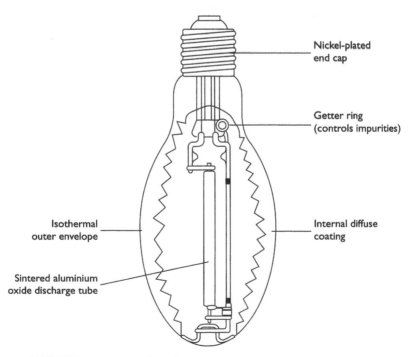

Figure 7.18 High pressure sodium lamp.

the lamp attains full luminous output. In some lamps a second discharge tube, identical to the main discharge tube, is incorporated which becomes energized following extinction of the first discharge tube, thereby reducing any time delay, before full luminous output is restored.

7.8.7 Induction lamps

The induction lamp, sometimes referred to as the 'electrodeless lamp', relies upon both magnetic and fluorescent principles for its operation. The constructional features of the induction lamps are shown in Figure 7.19.

Energy transfer using magnetism (in a similar manner to that of the electrical transformer) is initially employed, with the low pressure mercury filling in the lamp acting as a secondary coil of the transformer. The primary coil and a ferrite core are together referred to as the *antenna*. An alternating electrical current in the primary winding, typically at a frequency of 2.65 MHz., is supplied from an external source. The induced current in the mercury vapour, as a consequence of the magnetic field set up by the primary winding, produces emission of the ultraviolet photons which subsequently activate the phosphor coating of the outer glass envelope and in so doing produce radiation within the visible spectrum, in an identical manner to the production of light from a conventional fluorescent lamp.

The major benefit of the induction lamp is its extended lamp life, which is usually quoted as 60 000 hours. In the conventional lamp some of the electron-emitting oxides become detached from the electrodes upon starting and over a period of time these oxides become totally spent. With the induction lamp, however, there are no electrodes and so lamp life is greatly increased. Additional benefits of the induction lamp include:

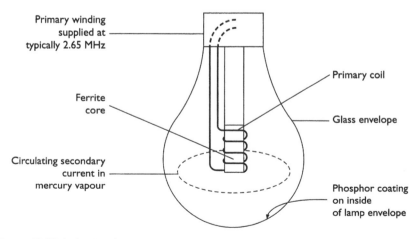

Figure 7.19 Induction lamp.

- increased luminous efficacy when compared to the conventional fluorescent lamp; and
- the light output is completely free of mains flicker and stroboscopic effects.

7.9 Lasers

The acronym laser refers to a device which produces 'light amplification by stimulated emission of radiation'. Typically optical radiation consists of a relatively large number of photons which are emitted spontaneously and additionally randomly in time. Such light is referred to as *incoherent light*. Conversely laser light is referred to as *coherent light* and the principle relies upon controlled stimulation of atoms and molecules so that individual photons are emitted 'in phase' with each other. A laser has the ability to produce a very intense and highly concentrated directional beam of electromagnetic radiation at a controlled and precisely defined wavelength.

Some classes of lasers are potentially extremely hazardous due to, inter alia, high levels of radiance, high energy (to which the human body can be exposed) and long distances over which the lasers can propagate.

The classification of lasers is covered by IEC 825-1[6] (IEC = International Electrotechnical Commission) and BS EN 60825-1 (1994).[7] Lasers are designated as Class 1, 2, 3A, 3B and 4.

By virtue of their low power or because of protective devices incorporated into the product, Class 1 lasers are safe under reasonable foreseeable conditions. Class 2 lasers are low power devices which emit visible radiation. Such devices are not intrinsically safe (safe within themselves) but eye protection is typically achieved naturally by an individual taking normal aversion responses, e.g. looking away. Lasers designated as Class 3A emit relatively high levels of radiation. If the beam of light is viewed directly, the power of the beam which will enter the eye will be not greater than that of the Class 2 laser. When the beam is viewed using some form of optical aid, the potential hazard becomes greater and more stringent safety precautions are required. Class 3B lasers are capable of causing damage to the eyes as a consequence of two fundamental causes:

- The output from the device is invisible and therefore an individual will not initiate normal aversion responses.
- The output beam from the device is so powerful that even before normal aversion responses are initiated, damage has occurred.

High power lasers designated as Class 3B are also capable of causing skin burns.

The most hazardous of all lasers are those designated as Class 4. These are extremely high powered devices which have the potential to cause instantaneous injury to both the eyes and the skin.

In some situations personal protective equipment (PPE) may be necessary. Safety eyewear used must be capable of operating efficiently at the wavelengths of radiation and power ratings typically encountered. Safety eyewear used in the protection against the effects of laser radiation is covered by, inter alia:

- Personal Protective Equipment at Work Regulations 1992[8]
- Personal Protective Equipment at Work: Guidance on the Regulations (L25)[9]
- Provision and Use of Work Equipment Regulations 1998[10]
- Guidance on the Provision and Use of Work Equipment Regulations (L22)[11]
- BS EN 207[12]
- BS EN 208.[13]

7.9.1 Display laser lighting installations

The majority of lasers used in display lighting have outputs which are sufficiently high to give rise to substantial eye injury. The eye is particularly vulnerable by virtue of the way it focuses the special light output produced by lasers which is coherent light and when viewed appears to have originated from a distant source. It follows that the image subsequently formed on the retina as a result of viewing a laser source is extremely small and therefore has a very high power density.

Lasers with a power rating of a few milliwatts are capable of causing damage to the retina in the short time duration before natural aversion responses are initiated. Natural aversion responses which include blinking, squinting and movement of the head can take typically up to 25 milliseconds. Furthermore, if the output from the laser is greater than 500 mW, then skin burns can also occur.

Reflection hazards from laser displays can also cause problems and surfaces in the vicinity of the display should be inspected and assessed for their potential to cause additional problems.

Types of lasers used in display techniques include:

- Argon – producing a blue-green output and used where high powers are required and often used in discotheques.
- Dye lasers – often used in conjunction with argon lasers.
- Helium-neon – producing a red light output, typically up to 10 mW, and often used in conjunction with scanning optics in order to produce 'writing' effects.
- Krypton – typically producing red but can also produce yellow and green light. Often used together with, or instead of, argon lasers where high power is required to be transmitted over long distances.

Further information in respect of display laser lighting can be found in 'The Radiation Safety of Lasers used for Display Purposes'.[14]

7.10 Optical fibres

Optical fibres are strands of glass, typically hair-thin, which are designed to transmit light rays along the axis of the fibre. They are often used in conjunction with light-emitting diodes (LEDs), where electrical signals are converted into optical signals which are then subsequently transmitted through an inner cylindrical core. The outer cladding of the optical fibre cable has lower refractive properties than those applying in the centre, which allows signals to be transmitted along the central core using the principle of *internal reflection*. The hazards likely to be encountered when transmitting laser radiation through optical fibre cable are included in BS EN 60825-2.[15]

7.11 Luminescence

Luminescence is the term used to describe emission of light by means other than the temperature radiation of a hot body. It can develop following:

* electrical emission;
* chemical reaction;
* living organisms; or
* the action of light or similar radiation.

A body can be luminescent as a consequence of *phosphorescence* or *fluorescence*.

The original discharge in a fluorescent or low pressure mercury vapour lamp emanates at two major wavelengths, i.e. 254 nm, 185 nm. A phosphor coating on the inner surface of the lamp envelope converts the original ultraviolet discharge wavelengths into visible light, the colour of which is significantly influenced by the phosphor composition.

Similarly a high pressure mercury vapour lamp, as discussed in Section 7.8.3, has a phosphor coating which increases the conversion of radiation from the original ultraviolet discharge into visible light and also assists in improving the colour properties of the lamp output. With fluorescence the light is no longer emitted when the initiating source is removed.

Other substances use fluorescence to advantage. When fluorescein is added to water in a container the transmission of light rays through the water is revealed by the vivid green fluorescence of the dye. Paraffin when placed under the light from a mercury vapour lamp will fluoresce, as will quinine (used as a febrifuge), bones and teeth.

Sodium fluorescein can be used by optometrists and/or ophthalmologists in connection with:

* checking intraocular pressure in the investigation of glaucoma;
* checking for aqueous leaks following eye surgery – the leaking fluid will absorb the sodium fluorescein and appear green;

- checking the fit of contact lenses;
- detection of corneal lesions; and
- assessing circulation in the retina – sodium fluorescein is injected into a vein and travels through the blood system. Pictures are taken as the dye circulates in the retina, a procedure which is referred to as *angiography*.

Furthermore fluorescence is useful in identifying and authenticating substances of value. Diamonds when placed under an ultraviolet light source will fluoresce vividly. Conversely imitations, which may appear similar under normal lighting conditions, do not fluoresce and so detection of imitations and forgeries is greatly increased.

With phosphorescence there is a corresponding afterglow associated with the light production. The property of phosphorescence has previously been used to advantage in the manufacture of luminous paints, since paints containing phosphorescent materials, when illuminated by sunlight (which contains ultraviolet wavelengths) would continue to glow. Developments in technology led to the production of luminous paints using some form of radioactive material, e.g. radium, thorium. The radiation from these substances, when acting on other substances contained in the paint, will produce visible radiation. The light emitted from such paints will continue indefinitely and there is no requirement for periodic re-exposure to sunlight.

Glow worms and fireflies glow in the dark, light being produced on the surface of their bodies as a consequence of oxidizing secretions. The radiation emitted by glow worms is at a wavelength very close to 555 nm, which corresponds to the wavelength of peak sensitivity of the photopic eye.

7.12 Lamp characteristics

Lamp characteristics include efficacy, lamp life, applications and run-up efficiency.

7.12.1 Lamp designations

Electric lamps are assigned designations. Table 7.2 shows the designations previously used together with the ILCOS (International Lamp Coding System)[16] designation which was introduced in 1993 by the International Electrotechnical Commission (IEC).

7.12.2 Lamp efficacy

Lamp efficacy can be considered as a 'value for money' indicator which gives details of the luminous output from a lamp and the corresponding electrical power required to provide the output.

Table 7.2 Lamp characteristics

Lamp type	Previous coding	ILCOS coding	Lamp efficacy (lumens per watt)	Quoted lamp life (hours)	Typical applications
Tungsten filament	GLS	I	10 to 18	1000 to 2000	Domestic
Tungsten halogen	TH	HS	15 to 25	2000 to 4000	Projector, car headlamps, display, stage lighting, traffic signals
High pressure mercury	MBF	QE	30 to 60	14 000 to 25 000	Car parks, older road lighting installations
Low pressure mercury (fluorescent)	MCF	FD (tubular)	65 to 95	6000 to 15 000	Offices, hotels restaurants, domestic
		FS (compact)	65 to 95	8000 to 10 000	
Metal halide	MBI	M	65 to 85	6000 to 13 000	Sports floodlighting, commercial interiors
Low pressure sodium	SOX	LS	70 to 150	11 000 to 22 000	Car parks, road lighting
High pressure Sodium	SON	S	55 to 120	12 000 to 26 000	Road lighting, car parks, civic areas
Induction		XF	70 to 80	60 000	Domestic, commercial, industrial

By definition:

$$\text{Luminous efficacy} = \frac{\text{Luminous flux output in lumens}}{\text{Electrical power input in watts}} \qquad (7.4)$$

Typical lamp efficacy values are given in Table 7.2.

7.12.3 Lamp life

It is impossible to specify precise figures for the life of a lamp although from laboratory work and field testing it is possible to predict the light output from a lamp in terms of the hours burned from new.

Two terms are used in connection with the output from lamps, i.e. *lamp survival* and *lumen maintenance*. Lamps incorporating a tungsten filament will fail completely before the deterioration in lumen output is thought to be significant. In such situations the term *average life* is used.

For other artificial light sources, typically discharge sources, the lumen output from the lamp will deteriorate throughout the life of the lamp before the point is reached where the lamp fails completely. In such cases lumen maintenance curves are used. Figure 7.20 shows typical lamp survival and lumen maintenance curves. The example used is that of the low pressure sodium lamp, similar curves are available for other types of lamps.

Lamp life can be affected by several factors including number of switchings, operating voltage, operating temperature and the presence of vibrations. Lamp manufacturers usually base the lamp life figures for hot cathode fluorescent lamps on eight switchings per day. For cold cathode fluorescent lamps the life is relatively unaffected by the rate of switching.

Typical lamp life values, quoted by lamp manufacturers, are given in Table 7.2.

7.12.4 Lamp applications

Table 7.2 gives details of typical applications of lamps.

7.12.5 Lamp run-up efficiency

A method for calculating the 'run-up efficiency', which describes the efficiency with which a tubular fluorescent lamp attains a maximum steady-state luminous output from a 'cold' start, was devised by Smith,[4] and the method can be used for any discharge lamp which takes time to reach maximum luminous stability. Figure 7.21 shows a plot of the variation in light output (in terms of the percentage of maximum attainable light output) with variation in time, from the time the lamp is switched on. Area 'A' represents the mathematical product of light

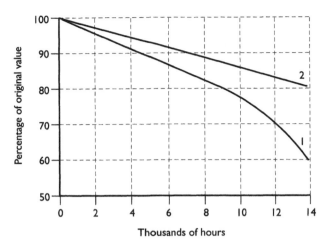

Figure 7.20 Typical lamp survival (1) and lumen maintenance (2) curves. Characteristics shown are typically those representing a low pressure sodium lamp.

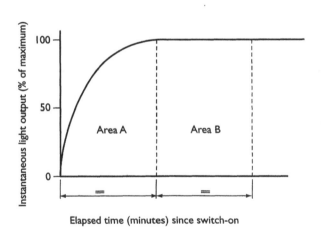

Figure 7.21 Concept of run-up efficiency.

output and elapsed time during the 'run-up' period (measured in percentage-minutes) and area 'B' represents the mathematical product of the steady-state light output and time (also measured in percentage-minutes) over the same time duration as that taken for the lamp to 'run-up', i.e. the same time duration as that applying for area 'A'.

The 'run-up efficiency' is then calculated from:

$$\text{Run-up efficiency} = \frac{\text{Area A}}{\text{Area B}} \times 100\% \tag{7.5}$$

7.13 Blue light hazard

Blue-light photochemical injury to the retina of the eye gives an indication of the potential retinal photochemical (blue light) hazard of a light source when viewed directly. It is referred to as a blue light hazard since the occurrence is mainly associated with wavelengths between 400 nm and 550 nm, i.e. in the blue region of the visible spectrum. In practice, the luminance of electric lamps is usually so high that an individual is unlikely to attempt to view them directly. It follows that if natural aversion responses are evoked, i.e. an individual will blink and/or look away, then the blue light hazard is unlikely to present major problems.

Blue light hazard is also important in the case of aphake individuals (those who have had a natural lens removed) and who have not been fitted with an ultraviolet (and blue light) absorbing intraocular lens.

7.14 Lamps used for other than normal lighting applications

Whilst the prime function of most lamps is to provide suitable and sufficient lighting for visual tasks to be undertaken safely and efficiently, some lamps are used for special purposes.

One such alternative use is in phototherapy, examples of which include the treatment of both neonatal jaundice and seasonal affective disorder (SAD). The liver acts as a human 'chemical exchange' system with important vital functions. It plays a significant role in the metabolism of fat, protein and carbohydrate and is also involved, inter alia, in processes relative to the absorption and storage of vitamins. The liver also excretes the breakdown products of haemoglobin which are the principle constituents of bile.

Haemoglobin in red blood cells is broken down so as to form bilirubin. In very young babies increased amounts of water-soluble bilirubin are present in the blood stream, usually as a result of the shorter life of the red blood cells combined with the condition of the infant's liver which is relatively immature. Under these circumstances, the bilirubin is not converted, at a fast enough rate, into a water-soluble derivative for subsequent secretion via the urine. It will therefore accumulate in the body and result in a yellowish tint to the skin. Treatment involves subjecting the patient to 'blue light'. Bilirubin absorbs visible light in the wavelength range 450 nm to 500 nm. It is then oxidized into non-toxic products and discharged.

The light detected by the retina of an individual, in addition to leading to visual processes, also stimulates the pineal gland in the brain. This gland produces a hormone called melatonin which, in humans, has a 24-hour secretion pattern. At night the pineal gland secretes melatonin but during the bright daylight the amount secreted is reduced. It is believed that many of the human body's circadian rhythms of activity are influenced by the melatonin secreted. Light therefore seems to be an important environmental factor in controlling 'activity-inactivity' patterns. If the body's clock becomes out of synchronization with the prevailing daylight and darkness then lethargy can develop which can, in some situations, be debilitating. It will be apparent that such situations can be exacerbated during the months of the year when daylight is at its minimum, i.e. in the winter months. The continued secretion of melatonin during daylight hours is thought to contribute to the symptoms shown by those suffering from seasonal affective disorder (SAD) which is a form of clinical depression. Exposure to lighting which emulates the output of a full spectrum can be effective in reducing the adverse effects of seasonal affective disorder.

Special lamps are available which are designed to emit predominantly in the ultraviolet range of the electromagnetic spectrum. Their uses include:

- non-destructive testing (NDT) in conjunction with fluorescent powders;
- validity check on bank notes and important documentation;
- special effects in advertising; and
- discotheque lighting.

Black light fluorescent lamps are manufactured with a special Wood's glass. This is a special deep violet-blue material which contains oxides of cobalt and nickel. The envelope appears black when unlit and almost totally absorbs the visible light emitted by the mercury discharge. Wood's glass is relatively transparent to long wave ultraviolet radiation and allows radiation at a wavelength of approximately 365 nm (generated by the discharge) to pass freely. When such lamps are viewed directly it can lead to misty and/or blurred vision as a consequence of fluorescence in certain parts of the eye. Such lamps should therefore be shielded from view.

Medical treatment involving lamps can include the use of UV lamps. For diagnostic applications UVA sources of radiation are often used. The nature of the ailment and the type of treatment required will dictate the required exposure to the source. UV lamps can additionally be used in the treatment of dermatitis.

Germicidal lamps, sometimes referred to as bacterial lamps, use ultraviolet sources whose main output wavelengths fall into the range 250 nm to 265 nm. This range gives optimum disinfection and sterilization characteristics and corresponds favourably with the maximum DNA absorption spectrum. Typically low pressure mercury discharge sources are used whose major spectral output line occurs at 254 nm, a wavelength which kills bacteria and other micro-organisms. Small amounts of radiation are emitted at a wavelength of 185 nm, producing small amounts of ozone, which is a deodorant and which, in the presence of water vapour, is bacterial and fungicidal. Germicidal lamps find extensive use in hospitals, including the sterilization of air, gases, liquids, etc., and also in breweries and food processing industries.

Ultraviolet lamps are used in many other workplace activities, including the photochemical curing of inks, plastics, paints and the treatment of certain medical conditions. Electromagnetic radiation of wavelengths between 200 nm and 400 nm is referred to as *actinic* since it brings about chemical changes. The effects of the radiation varies in accordance with their penetration through the human skin. Wavelengths in the range 300 nm to 400 nm are known as *biotic* and stimulate living tissues whereas those wavelengths in the range 200 nm to 300 nm are referred to as *abiotic*, and are inimical to life.

Ultraviolet lamps are also used in insect traps, the underlying principle being that insects have eyes which are sensitive to UV radiation. When attracted to the UV source the insects make contact with a grid which carries an electrical potential which, though relatively small, is nevertheless lethal to the insects.

Consider the curing of dental fillings. An open beam dichroic mirror lamp, selected for its concentrated beam and constancy of luminous output will, when using a halogen lamp, emit a range of wavelengths from ultraviolet through the visible band to infrared. Many of the materials used in dental restoration are, however, only hardened by wavelengths of radiation in the blue and violet bands of the spectrum.

Some of the compounds used in early attempts at curing used UV light in order to effect a speedy curing. A problem existed, however, whereby dental staff were being subjected to UV reflected from patients' teeth.

Fortunately there are some benefits to using visible light for curing:

• It has a greater curing depth than ultraviolet light and so therefore it allows the filling compound to be used for both posterior and shallow anterior fillings.
• It allows a visible indication of the location of the curing.
• The correspondingly reduced curing time is welcomed by patients.

By using suitable filters, the wavelengths of near ultraviolet together with some of the lower energy visible wavelengths are blocked from the normal emission spectrum.

In practice the physical size of the tungsten halogen lamp used in conjunction with fibre optics allows the production of an instrument which is slender in construction which in turn allows good directional control, simultaneously leaving the dentist with a relatively unobstructed view of the dental cavity.

Sun tanning salons use lamps which often have their main output in the UVA range. Clients and staff must take suitable precautions so as to avoid over exposure to the effects of such lamps.

Infrared lamps emit electromagnetic radiation predominantly in the IRA range and tend to find uses in heat treatment, paint drying and associated processes. Additionally medical uses for IR lamps include therapeutic and diagnostic applications.

7.15 Standard illuminants

Surface colours will appear different when viewed under light sources with different spectral characteristics. When striving for accuracy when

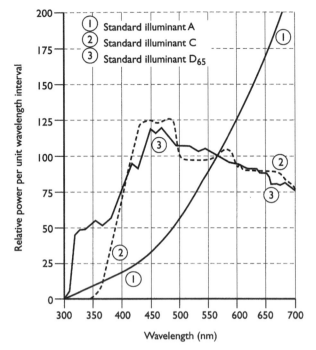

Figure 7.22 Spectral power distribution of standard illuminants.

measuring colours it is advantageous to specify the spectral characteristics of several different types of illuminants. Figure 7.22 shows the relative spectral power distributions of three standard illuminants, i.e. A, C and D_{65}.

- **Standard illuminant A:** The spectral power distribution of a GLS incandescent lamp is not dissimilar to that of the full radiator. Standard illuminant A is therefore defined as an incandescent source, with a colour temperature of 2856 Kelvin.
- **Standard illuminant C:** This equates to average daylight, not including the ultraviolet region, with a correlated colour temperature of 6774 Kelvin.
- **Standard illuminant D_{65}:** This equates to average daylight, which includes the ultraviolet region, with a correlated colour temperature of 6504 Kelvin.

7.16 Beta lamps

Beta lamps are self-illuminating light sources which have the advantage that they require no mains electrical supply for their operation. They are used in areas where the presence of an explosive mixture of gases or vapours is present or is likely, e.g. petrochemical industries and marine and aviation applications. The light output produced is relatively weak but nevertheless the advantages of self-power outweigh the many obvious disadvantages of using mains powered equipment in hazardous environments.

Beta lamps are essentially sealed borosilicate glass capsules which are internally coated with phosphor and filled with tritium. Tritium is a radioactive isotope of hydrogen having a half-life of 12.4 years. Figure 7.23 shows the constructional features of the lamp.

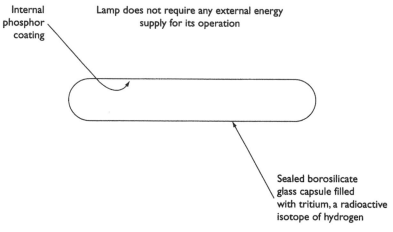

Figure 7.23 Beta lamp.

7.17 Tyndall beam lamp

The measurement of airborne dust is an important factor in determining air quality within an interior. Many dust particles are of such small size that they remain undetected by the naked eye. If, however, in such circumstances a beam of light is directed towards a cloud of dust, the particles will reflect the light and the dust becomes readily visible to an observer.

The Tyndall beam lamp, named after J. Tyndall, and often referred to affectionately as the 'dust lamp', makes use of this basic phenomenon to allow observation of the behaviour of clouds of dust particles.

Typically a portable luminaire contains a reflector which is designed so that the parallel beam of light emitted by the lamp penetrates the environment which is suspected of being dusty. An observer will then be able to detect the presence of any dusts. The general arrangement is as shown in Figure 7.24.

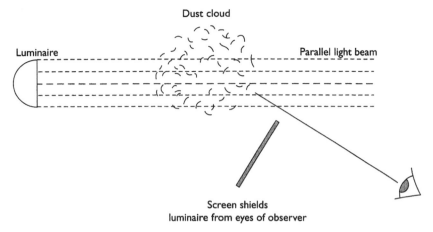

Figure 7.24 Concept of the Tyndall beam lamp.

7.18 Handling of new lamps and the disposal of spent lamps

New or in-service tungsten halogen lamps should not be touched with bare hands. Oils and greases, both of which are present on the finger tips, may be passed on to the surface of the quartz envelope and during operation of the lamp, blistering is likely to occur leading to premature lamp failure.

When such lamps have reached the end of their useful life, they should be broken in a strong container so that the fragments of the quartz envelope can be contained. The gas fill in tungsten halogen lamps has a slight toxic content and as a consequence large amounts of such lamps must therefore be broken in an environment which is well-ventilated.

Lamps contain toxic substances which may include cadmium, mercury and lead. Cadmium does not degrade and it cannot be destroyed. The pollution due to cadmium presents a substantial problem since it is a substance affecting all sectors of the environment. Cadmium is harmful to humans and also to some aquatic species and in addition it is capable of migrating through several environmental media.

If cadmium is not correctly deposited in a secure waste disposal site it can become deposited in the soil from where it can be absorbed by plants and subsequently enter the food chain. Once the cadmium has been absorbed it can lead to liver, kidney and brain damage.

Fundamentally mercury cannot be destroyed and can persist in mud and eventually pollute water courses. Mercury readily vaporizes in cold temperatures and at a faster rate in higher temperatures. Individuals can be contaminated by mercury inhalation or by skin contact when it can manifest itself as skin rashes. When it is absorbed into the body it lodges in the kidneys and the liver.

Sodium lamps are both fire and explosion hazards until broken correctly and the sodium neutralized. Additionally the storing of sodium lamps is hazardous due to the hygroscopic nature of sodium, i.e. it absorbs moisture from the air. In damp conditions sodium can absorb moisture in sufficient quantities to allow the lamp to heat up and subsequently to ignite. If this is allowed to happen then the chemical reaction of water with sodium produces sodium hydroxide, which is corrosive, and hydrogen which is flammable. The reaction is so fierce that significant quantities of heat are generated. The energy requirement for the ignition of hydrogen is approximately 20×10^{-6} Joules. By volume a 4 per cent mixture of hydrogen and air produces flammable conditions. Ignition of a mixture of flammable gas and air of significant volume produces an explosion.

Fluorescent lamps, which are low pressure mercury lamps, release powders when broken and such powders will have been contaminated with mercury during operation of the lamp. The inhalation of the powder should be prevented.

Mercury produces a vapour at ambient temperatures. Broken lamps will therefore release mercury vapour and liquid mercury although in relatively small quantities. In some cases the mercury may be converted by anaerobic decomposition into methyl mercury, which is extremely toxic.

Lamp disposal must be carried out in accordance with current legislation, including:

- Control of Substances Hazardous to Health Regulations 1994;[17]
- Control of Substances Hazardous to Health and Control of Carcinogenic Substances: Control of Substances Hazardous to Health Regulations 1994 Approved Codes of Practice (L5);[18] and
- Special Waste Regulations 1996 (as amended).[19]

See Appendix (A.3.11 and A.3.12).

Information on lamp disposal may be obtained in the UK from the Environment Agency.[20]

7.19 Ultraviolet radiation from tungsten halogen lamps

The construction and operation of a tungsten halogen lamp is described in Section 7.7.2. The lamp envelope of a tungsten halogen lamp is constructed from quartz, because of its thermal properties. The combination of higher filament temperatures which are typically in the range 3000 Kelvin to 3400 Kelvin, together with the quartz envelope which is a good transmitter of ultraviolet radiation, gives rise to a greater emission of ultraviolet radiation from tungsten halogen lamps when compared with the conventional tungsten filament lamps. Tungsten filament lamps operate at filament temperatures of typically between 2800 Kelvin and 3000 Kelvin.

High powered lamps which are typically used for theatres and studios must be installed in luminaires complying with the requirements of current legislation. They should have a broken lamp shield and an ultraviolet shield fitted. Low powered lamps of the capsule or linear type, without integral outer glass envelopes, should normally be used in luminaires which are shielded.

7.20 Ultraviolet radiation from high intensity discharge lamps

Some high pressure mercury vapour and metal halide lamps will emit ultraviolet radiation during either starting, warm-up or normal running conditions. It is essential therefore that they are operated in luminaires specially designed for their use, e.g. those which contain a safety shield.

7.21 Effects of operating broken mercury vapour and metal halide lamps

High intensity mercury vapour and metal halide lamps present a danger if they are allowed to continue to operate when the outer glass envelope of the lamp is broken. The gas-containing discharge tube uses quartz which freely transmits the ultraviolet radiation produced. The discharge tube is surrounded by an envelope of borosilicate glass which, under normal operating conditions, absorbs the invisible ultraviolet radiation and bluish visible light emitted (see Figure 7.25).

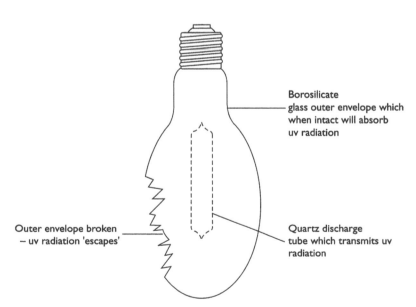

Borosilicate glass outer envelope which when intact will absorb uv radiation

Outer envelope broken – uv radiation 'escapes'

Quartz discharge tube which transmits uv radiation

Figure 7.25 Effects of operating metal halide and high pressure mercury vapour lamps with broken outer envelope.

An individual exposed to ultraviolet radiation similar to that emitted by a broken mercury vapour or metal halide lamp would not necessarily feel any immediate effects but may, within a few hours, start to feel a painful reddening of the skin known as *erythema*. Furthermore, the exposed individual may be aware of a 'gritty' or 'sandy' sensation in the eyes which is referred to as *photokeratitis*.

It follows that any damaged lamps which are still capable of operating should be taken out of service immediately and disposed of in accordance with current standards, as discussed in Section 7.18.

Chapter 8

Luminaires

8.1 Introduction

The luminaire, formerly referred to as the 'light fitting', provides support, protection and means of electrical connection to the lamp which is contained within the luminaire. Additionally the luminaire has to be able to operate safely and to withstand the environmental conditions in which it is likely to be installed.

Luminaires may also be classified according to the following categories:

- type of protection provided against electric shock;
- the degree of protection provided against the ingress of dust or moisture;
- 'type' protection provided against electrical explosions; and
- according to the characteristics of the material of the surface to which the luminaire can be fixed.

Those luminaires which are designed for use internally, and those for use in some external applications, will normally operate efficiently in dry and well-ventilated atmospheres.

Some luminaires are installed in atmospheres which are far less acceptable. In order to be able to perform within the environments likely to be encountered, the equipment must therefore be manufactured so as to comply with strict specifications. Such environments are referred to as *hostile* or *hazardous*. Examples of hostile areas include:

- high humidity conditions, requiring drip-proof equipment;
- dusty and corrosive atmospheres;
- food factories, where the interior walls, floors and ceilings are required to be hosed down.

Hazardous atmospheres are usually created by the presence of flammable or explosive dusts and gases in the atmosphere.

8.2 Optical control of light output from luminaires

When a bare lamp is used, i.e. one without any form of control of the directional qualities of the light emitted, then the distribution of light is likely to be completely unacceptable. Furthermore the bare lamp is also likely to create a source of disability glare to the occupants of the interior. It is likely that the lighting installation will be uneconomical and whilst some fraction of the light output from the bare lamp will reach the working plane either directly or indirectly, the efficiency of the installation will be relatively low.

It will be clear that some means of control of the light output from the bare lamp is essential and four of the most widely-used methods are detailed:

- **Obstruction**: When a bare lamp is installed within an opaque enclosure, which has only one aperture from which the light can escape, then the light distribution from the basic luminaire will be severely limited (Figure 8.1).
- **Reflection**: This method of light control uses reflective surfaces, which may range from matt to specular. By comparison with the 'obstruction' method of light control, reflection is more efficient. With reflection, stray light is collected by the reflectors and then redirected (Figure 8.2).
- **Diffusion**: When a lamp enclosure is constructed of a translucent material, two benefits will materialize. The apparent size of the light source is increased and simultaneously there is a reduction in perceived brightness. One disadvantage as a consequence of the use of diffusers is that they absorb some of the light emitted from the source itself and so therefore there will be a reduction in the overall luminaire efficiency. Figure 8.3 shows the concept of diffusion.

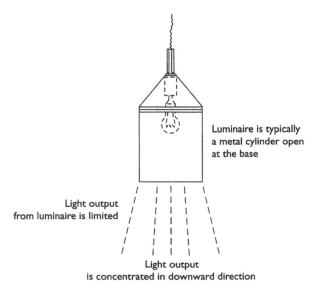

Luminaire is typically a metal cylinder open at the base

Light output from luminaire is limited

Light output is concentrated in downward direction

Figure 8.1 Light control by obstruction.

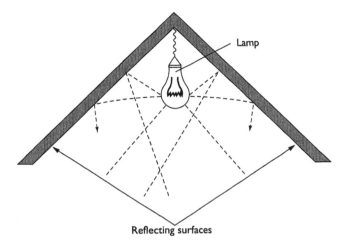

Figure 8.2 Light control by reflection.

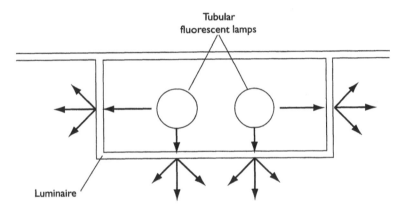

Figure 8.3 Light control by diffusion.

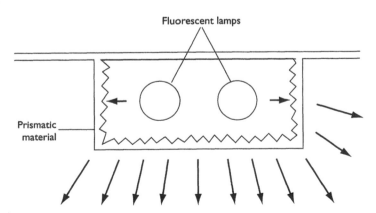

Figure 8.4 Light control by refraction.

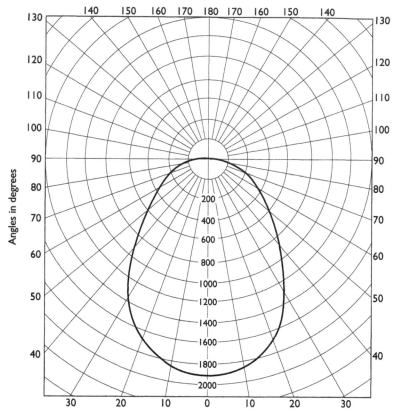

Figure 8.5 Typical polar diagram. Radial lines represent values of luminous intensity (candela).

- **Refraction**: This method utilizes the prism effect so as to 'bend' the light output from the bare lamp in the required direction. This method of light control has the benefits of effective glare control and additionally of producing an acceptable level of luminaire efficiency (Figure 8.4).

Prisms which are constructed from either glass or plastic can be designed so as to control the light output from the bare lamp. Plastics are used frequently in prismatic controllers.

8.3 Distribution of light from luminaires

Light distribution from luminaires can be represented graphically using polar diagrams when the intensity distribution is symmetrical about the vertical axis through the luminaire. Isocandela diagrams are used to present the information when the luminaires have a non-symmetrical intensity distribution.

8.3.1 Polar diagram

Where the distribution of luminous intensity is identical through all vertical planes a polar diagram is used. The radial lines emanating from the centre of the diagram, which coincides with the centre of the light source under investigation, represent the magnitude of the luminous intensity of the source measured in candela. The magnitude, at any given angle to the vertical through the luminaire, can therefore be read directly or interpolated from the diagram. Figure 8.5 shows a typical polar diagram. Manufacturers of fluorescent luminaires frequently use a form of 'split' polar diagram which shows details of the intensity distribution from a luminaire relative to the two major axes through the luminaire centre, i.e. *axial* and *transverse*.

8.3.2 Isocandela diagram

When the distribution of luminous intensity is not symmetrical about the vertical axis through the luminaire, the polar diagram cannot accurately represent the appropriate information. For such luminaires, the isocandela diagram often affectionately referred to as an 'onion diagram' is used. The diagram, which is graduated in angles of azimuth and elevation, conveys luminous intensity distribution information for a multitude of planes. The magnitude of the luminous intensity, in candela, for any direction of view from the centre of the light source can be found from the diagram. If the required direction, with respect to the centre of the light source, is such that it does not exactly coincide with a given isocandela contour, then the corresponding value of luminous intensity can be interpolated. Figure 8. 6 shows a typical isocandela diagram.

8.3.3 Isolux diagram

Polar diagrams and isocandela diagrams, which were considered in Sections 8.3.1 and 8.3.2, show luminous intensity values of the light distribution from luminaires. It is, however, more useful to be able to predict the illuminance, on the working plane, due to a given luminaire. Luminaire manufacturers therefore publish isolux diagrams, which show illuminance contours on the working plane due to the light output from a particular luminaire. Figure 8.7 shows a typical isolux diagram. It is possible to apply correction factors to the information shown on the diagram for any variation in the mounting height of the luminaire above a particular working plane.

8.4 Luminaires for hazardous and hostile environments

Luminaires used in hazardous and hostile environments must be manufactured to stringent conditions so that they are capable of operating safely

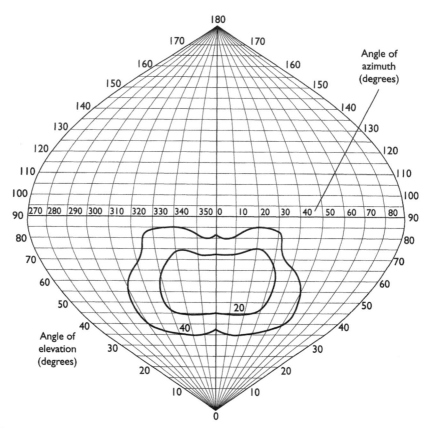

Figure 8.6 Typical isocandela diagram. Figures on contour lines represent luminous intensities in candela per 1000 lamp lumens.

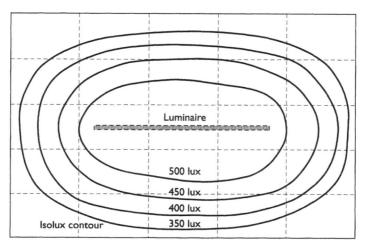

Figure 8.7 Typical isolux diagram. Plan view of working plane showing isolux contours.

and efficiently in the atmospheres in which they are likely to be installed. Classification of luminaires is considered in relevant British Standards as detailed in Section 8.5.

8.5 Classification of luminaires according to relevant British Standards

Luminaire classification is included in the following British Standards; BS 4533,[21] BS 5345[22] and BS 6467.[23]

8.5.1 BS 4533[21]

BS 4533[21] is the British Standard relative to luminaires. (see also EN 60-598[24]). Details are given of the protection provided by luminaires against electric shock, ingress protection and the material of the supporting surface.

- **Electric shock**: Equipment is designated as Class 0, I, II, or III depending primarily upon the degree of electrical insulation provided.
- **Ingress protection**: The ingress protection (IP) system of luminaire classification is a two-digit classification indicating the magnitude of protection against the ingress of foreign objects and liquids. The first digit gives the protection provided against touching live or moving parts of the equipment and against solid objects entering the luminaire. The first digit ranges from '0' (zero) which represents no protection, to '6' which represents complete protection and which is referred to as dust-tight. If the first digit is '5' the luminaires are referred to as dust-proof. The second digit gives details of the protection afforded by the luminaire against the ingress of liquids under varying conditions. The second digit ranges from '0' to '8'. '0' (zero) refers to no protection, and the references and corresponding descriptions include: (2) drip-proof, (3) rain-proof, (4) splash-proof and (5) jet-proof. A dust-tight luminaire which is capable of being hosed down would therefore carry the IP65 designation. Some authorities have used a modification to this system where a third digit represents the level of protection afforded by the luminaire against mechanical impact damage.
- **Material of supporting surface**: BS 4533[21] includes information in respect of the allowable materials for the supporting surfaces of luminaires, e.g. flammable and non-combustible surfaces.

8.5.2 BS 5345[22]

BS 5345[22] is the British Standard Code of Practice which relates to the selection, installation and maintenance of electrical apparatus for use in potentially explosive atmospheres (other than mining or explosive

Table 8.1 Luminaire classification according to type of protection as detailed in BS 5345[22]

Type designation	Description	Comments
Ex 'd'	Flameproof enclosure	For use in Zone I areas
Ex 'p'	Pressurized enclosure	For use in Zones I and 2 areas
Ex 'e'	Increased safety	For use in Zone I and 2 areas
Ex 's'	Special protection	Allows the development of new techniques before the issue of standard specifications
Ex 'N'	Non-sparking	For use in Zone 2 areas

processing and manufacture). This standard categorizes hazardous areas into zones where:

- Zone 0 is an area where an explosive gas/air mixture is continuously present.
- Zone 1 is an area where an explosive gas/air mixture is likely to occur during normal operation.
- Zone 2 is an area where an ignitable concentration of gas/air is not likely to occur during normal operation and further that any such concentration will only exist for a short time duration.

Mains operated lighting is not permitted in Zone 0 areas, although beta lamps are permitted. Beta lamps are detailed in Section 7.16.

Type of protection considers the suitability of different luminaire enclosures for use in various environments, as shown in Table 8.1.

8.5.3 BS 6467[23]

BS 6467[23] is the British Standard entitled 'Electrical apparatus with protection by enclosure for use in the presence of combustible dusts'.

Many materials found in industrial situations may be in the form of a cloud of dust present in the atmosphere. It will be evident that if the cloud becomes ignited, an explosion may occur, and layers of dust which may be present on existing apparatus, may raise the temperature to a value where smouldering may commence. If this situation occurs then there is the likelihood of a fire hazard. If the dusts were metallic in nature then there is the additional problem of the possibility of breaking down electrical insulation with arcing leading ultimately to short circuits.

8.6 Downlighter luminaires

Those luminaires classified as downlighters will emit the major proportion of their light output in the downward direction. They are typically ceiling-mounted or ceiling-suspended. When compared with uplighters, downlighters are far more efficient since they provide direct lighting of the working plane, and do not rely totally on light reflected from room surfaces.

8.7 Air handling luminaires

An air handling luminaire is one through which stale exhaust air is taken from an interior. In general terms the luminaire fills a dual role in that it provides a light output for the benefit of the occupants of the interior and also it replaces an exhaust grille in a ventilating and/or air conditioning system.

Heat is generated by the process of conversion from electrical energy to light energy and also by the normal operation of the control gear which is an essential part of the discharge lamp circuit. It is beneficial to remove hot air from the interior before it becomes part of the circulating air. It is important when using air handling luminaires which contain fluorescent lamps not to extract too much heat. The luminous output from fluorescent lamps is extremely temperature-dependent and too much hot air extraction will lead to a reduction in light output from the lamp(s). Figure 8.8 shows the general principle of operation of an air handling luminaire.

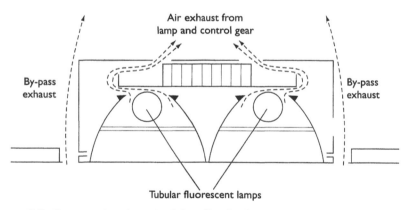

Figure 8.8 Cross-section through air-handling luminaire.

8.8 Uplighter luminaires

Those luminaires which are classified as uplighters will emit the major proportion of their light output in the upward direction. Figure 8.9 shows the general principle of lighting using uplighters. When compared to downlighters, uplighters are relatively inefficient since they

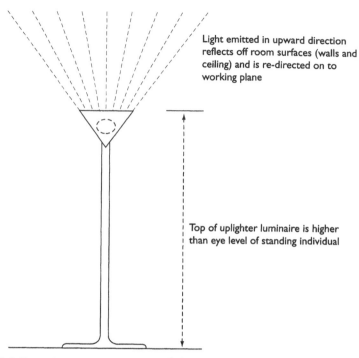

Light emitted in upward direction reflects off room surfaces (walls and ceiling) and is re-directed on to working plane

Top of uplighter luminaire is higher than eye level of standing individual

Figure 8.9 Principle of uplighter luminaire.

rely on light being reflected from room fabrics. This places a greater importance on the room decor and reflectance values. The limitations placed on the use of uplighters in interiors is considered in detail in Section 9.7.

8.9 Luminaire materials

The materials from which luminaires are constructed typically include metals, glass and plastic. Glass had been used extensively in earlier luminaires but has largely been replaced by plastics, except for some specialist applications, e.g. where the light source produces excesses of heat, in which case glass is preferred from a safety viewpoint. Furthermore, glass still finds use in situations where there is the likelihood of physical damage to the luminaire, since glass will be more robust.

Two types of glass are in common use for the manufacture of the bowls of luminaires, i.e. *borosilicate glass*, which can withstand high working temperatures, and *soda-lime glass*. Both of these types of glass have to be subjected to toughening processes in order to provide additional protection if they are likely to be installed and operated in environments where there are rapid temperature variations.

Plastics, which are either *thermosetting* or *thermoplastic*, find their main use in the construction of diffusers. Polystyrene, acrylic and polycarbonate are used for diffusers. Polycarbonate, which is used in the manufacture of some types of safety spectacles, has a greater impact resistance than acrylic although both materials have been used for the construction of enclosures for use in areas where the luminaires are subjected to acts of vandalism. A form of plastic known as *glass-reinforced plastic* (GRP) is often used for luminaire canopies.

Sheet steel and aluminium are often used in the construction of luminaires, primarily for the construction of the main body of the luminaire, but also, on occasions, for the louvres which are included in order to restrict glare.

8.10 Mechanical strength of luminaires

Mechanical characteristics of luminaires include strength, windage resistance and resistance to vibration.

Luminaires must be capable of operating safely and efficiently within the environments in which they are expected to be installed. If they are operating in external environments they are likely to be subjected to meteorological factors such as wind, ice and snow, all of which must be taken into account when considering the construction of the luminaires. Luminaires subjected to strong winds may also suffer the effects of vibrational oscillations which may lead to adverse operation.

8.11 Control gear

The term is used affectionately to describe the equipment which is necessary for the safe and efficient operation of discharge lamps. The function of the control gear is essentially twofold:

- It assists in providing a high voltage pulse to enable the arc to be established within the discharge tube in the lamp.
- Once the arc has been established, the control gear takes on the role of a current limiting device in order to restrict the current flowing through the lamp. Discharge lamps exhibit a *negative resistance characteristic*.

It has to be appreciated that heat is produced during normal operation of lamps and control gear. Nevertheless many items of control gear have recommended maximum temperatures and it is essential that such values are not exceeded.

Individual items of electrical and/or electronic equipment used include ballasts, transformers, ignitors and power factor correction capacitors. The items of control gear will consume power from the supply and the resulting electrical energy costs will have to be borne by the consumer. It is therefore beneficial to be aware of the power ratings of control gear when considering the economics of lighting.

8.12 Electromagnetic interference (EMI) and radio frequency interference (RFI)

The operation of lamps and associated control equipment must be such that they satisfy current regulations in respect of electromagnetic radiation interference (EMI) and radio frequency interference (RFI).

8.13 Polychlorinated biphenyls (PCBs)

In early versions of luminaires, the power factor correction capacitor within the control circuitry of the lamp used polychlorinated biphenyls (PCBs) as the dielectric or insulating material.

Accidental exposure of humans to PCBs is an area of great concern. In those exposed occupationally, a wide range of health effects may be experienced including:

- changes to the skin and mucous membrane;
- swelling of the eyelids and excessive eye discharges;
- burning sensation and oedema of the face and hands;
- chloracne;
- hyperpigmentation of the skin and thickening of the skin;
- discoloration of the fingernails;
- digestive symptoms including abdominal pain, nausea, vomiting and jaundice; and
- neurological symptoms including headaches, dizziness and depression.

Chapter 9

Lighting for interior applications

9.1 Introduction

Lighting within interiors includes lighting provided for offices (including display screen equipment areas), domestic premises and warehouses.

9.2 Office lighting

It is possible to classify the main requirements for optimum lighting conditions within interiors as:

- health and safety;
- visual performance;
- aesthetics; and
- personal comfort.

When striving to obtain an office environment which is both aesthetically pleasing and acceptable to the eye it is usual to include:

- an analysis of those visual tasks which are likely to be encountered within an office;
- the striking of a balance between the quantity of daylight entering an office interior and the corresponding requirements from artificial lighting;
- the striking of a balance between direct and diffuse lighting contributions so that, where practical, no adverse three-dimensional and/or modelling effects will be produced; and
- the provision of an environment which is free from both glare and any associated distractions.

When considering the health and safety of personnel, office lighting should ideally allow room occupants to carry out their normal duties in a manner which, under normal conditions, is considered to be safe. The lighting provided must not, under any circumstances, place any occupants of the room at risk. Furthermore it is necessary to provide for the safety of occupants in the event of an essential evacuation of the premises. In the United Kingdom emergency lighting is a requirement under The Building Regulations[25] and is covered by BS 5266 Emergency Lighting.[26]

Visual performance is the ability of the eye to carry out a particular visual task with both speed and accuracy. It is dependent upon the level of prevailing illuminance; the relationship is one that follows the law of diminishing returns.

The quality and quantity of lighting provided will significantly influence the visual environment subsequently produced, which is also dependent upon the details of the visual tasks undertaken in the offices.

It is important not to overlook both aesthetics and personal comfort when considering optimum office lighting conditions. There is a psychological influence on the occupants of an interior, caused by the decor and room appearance, and subsequently this is likely to have a 'knock on' effect upon worker behaviour which will ultimately affect output productivity.

9.3 Techniques and practice of interior lighting

It is possible to categorize interior lighting into general lighting, localized lighting or local lighting:

- **General lighting**: A general lighting system is one which attempts to provide a constant illuminance across a working plane in an interior. It is, however, extremely unlikely that a uniform level of illuminance will be produced at all points across a horizontal working plane.

  It has to be appreciated that general lighting does not take into account the visual tasks likely to be undertaken in an interior. This can have both advantages and disadvantages. One advantage is that it allows a degree of flexibility when locating individual areas within an interior where visual tasks can be carried out. A disadvantage of general lighting is that such systems can be very wasteful of energy since some areas within an interior are illuminated to a level greater than that required.

- **Localized lighting**: Localized lighting systems usually provide a required illuminance on the work areas in combination with a reduced level of illuminance in non-working areas, for example walkways. An often used example of this is found in open plan offices where workstations are lit using uplighters. Simultaneously, walkways and other non-work areas are lit by means of a number of ceiling-mounted downlighters. It will be evident that this system is correspondingly less wasteful of energy than the general lighting system.

- **Local lighting**: Local lighting systems provide illuminance over a relatively small area in which the visual task is located. It is often used

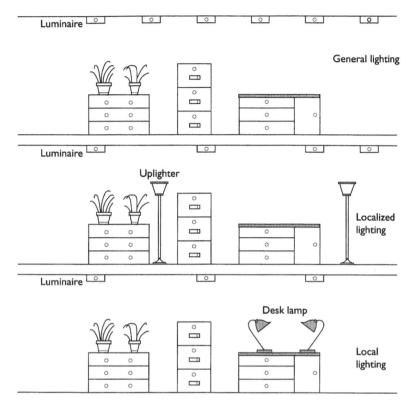

Figure 9.1 Types of lighting system

in combination with general lighting so that together the local lighting and general lighting will produce the required illuminance on, and surrounding, the visual tasks.

Figure 9.1 shows the differences between the types of lighting system described.

9.4 The lumen method of lighting design

The 'lumen method' of lighting design is almost invariably used as the basis for the calculation of the required number of luminaires when considering interior lighting requirements. The illuminance is found using this method from:

$$\text{Illuminance (E) (in lux)} = \frac{\text{Luminous flux (in lumens)} \times \text{Maintenance factor} \times \text{Utilization factor}}{\text{Surface area (in metres}^2)} \tag{9.1}$$

The luminous flux emitted by a lamp is usually quoted in manufacturers' catalogues, and, by way of an example, the output of a 1500 mm white tubular fluorescent lamp (26 mm diameter) is typically 4500 lumens.

The *maintenance factor* is a parameter which takes into consideration the gradual reduction in light output due to a combination of both (a) the number of hours the lamp has burned from new and (b) dirt, grime and generally foreign bodies appearing on the lamp and/or luminaire.

The *utilization factor* is a parameter which indicates the degree of light output from the luminaire(s) which is usefully available at the working plane, usually expressed as a fraction of the total light output from the luminaire(s). This figure is clearly influenced by the geometry of the interior under consideration and also the reflectance values of the room fabrics.

Worked example

An office is to be illuminated to a minimum design level illuminance of 500 lux, using 1500 mm tubular fluorescent lamps (26 mm dia.), each with a luminous output of 4500 lumens. The office measures 15 metres by 10 metres. The values of maintenance factor and utilization factor can be taken as 0.8 and 0.4 respectively. Determine the minimum number of lamps required and suggest a possible layout using single lamp luminaires.

Using the 'lumen method', consider the illuminance provided by one lamp:

$$\text{Illuminance (lux)} = \frac{\text{Luminous flux (lumens)} \times \text{Maintenance factor} \times \text{Utilization factor}}{\text{Surface area (metres}^2)}$$

$$\text{Illuminance} = \frac{4\,500 \times 0.8 \times 0.4}{150} = 9.6 \text{ lux}$$

It follows that the number of lamps required is (500 ÷ 9.6) = 52.08, a figure which is clearly unobtainable. To obtain a minimum design level of illuminance of 500 lux, it is evident that more rather than less lamps need to be employed. It is therefore recommended that 54 lamps be used, possibly in a 9 × 6 array, as shown in Figure 9.2.

9.5 Preferred illuminance and luminance ratios for interiors

Many lighting installations consist of a regular pattern of luminaires which in combination produce a flow of light which together with daylight admitted through side windows, gives suitable and sufficient lighting together with an acceptable level of modelling.

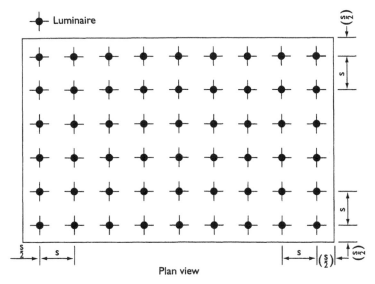

Figure 9.2 Suggested layout of luminaires in worked example. Centre of perimeter luminaires are spaced (½) from adjacent walls; other luminaires have equidistant spacing between centres (s).

It is, however, preferable to consider the lighting effects produced in an interior in terms of illuminance ratios and luminance ratios as shown in Figures 9.3 and 9.4.

9.6 Lighting for areas containing display screen equipment

In interiors which contain display screen equipment the objective, in respect of lighting, is simple: 'to obtain an acceptable balance of all light sources, both natural and artificial, which, with strategic positioning of the operator and the display screen (with respect to the light sources) will produce an environment which is free from both glare and veiling reflections'. The lighting of areas containing display screen equipment is considered in detail in Chapter 13.

9.7 Uplighter installations

The use of uplighters is becoming more prevalent in interiors and in addition it is a very effective method of preventing veiling reflections. The obvious advantage of uplighters is that the light source within the uplighter luminaire is not visible to the naked eye of an individual. Uplighters, of necessity, direct their light initially onto the room fabrics

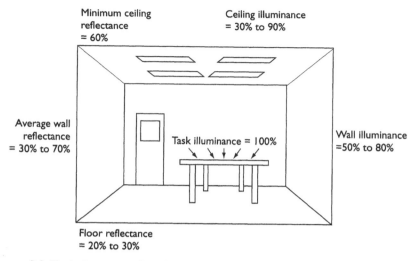

Minimum ceiling
reflectance
= 60%

Ceiling illuminance
= 30% to 90%

Average wall
reflectance
= 30% to 70%

Task illuminance = 100%

Wall illuminance
=50% to 80%

Floor reflectance
= 20% to 30%

Figure 9.3 Typical room surface illuminance ratios. Illuminance values quoted are relative to a task illuminance of 100 per cent.

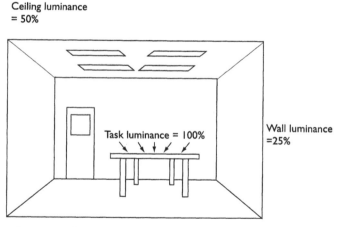

Ceiling luminance
= 50%

Task luminance = 100%

Wall luminance
=25%

Figure 9.4 Typical room surface luminance ratios. Luminance values quoted are relative to a task luminance of 100 per cent.

(walls and ceilings) and their operation is consequently less efficient when compared with ceiling-mounted luminaires which are often referred to as downlighters. This reduction in efficiency is often acceptable since the avoidance of veiling reflections is usually the main criterion. There are four generally accepted locations for uplighters, i.e. wall-mounted, ceiling-suspended, free-standing and furniture-mounted, all as shown in Figure 9.5.

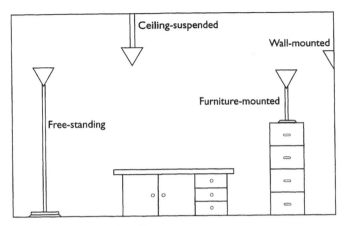

Figure 9.5 Uplighter locations in offices.

There are, however, some obvious disadvantages associated with the use of uplighters. The ultimate lighting conditions produced by an uplighter installation are dependent upon:

- room dimensions;
- room fabric colour finishes;
- room fabric reflectances and hence luminance values;
- room fabric specularity;
- lamp colour output; and
- lamp flicker.

It is important to appreciate that when considering an uplighter installation, the colour output of the light sources must be thought of as an inseparable combination with the colour finish of the ceiling fabric. The apparent colour of the ceiling is strongly influenced by:

- the colour output of the light emitted by the lamps; and
- the spectral reflectance characteristics of the ceiling fabric.

It will be evident that if a white ceiling is illuminated by lamps with a white output then the resulting room appearance will be white. If, however, a white ceiling is illuminated by lamps whose colour output is 'non-white' then the ceiling will appear to be the colour of the 'non-white' source. Conversely if a 'non-white' ceiling is illuminated by a white light source, the room appearance will be that of the colour of the ceiling itself.

Flickering light sources are at least annoying and in many cases extremely distracting and often with uplighters the situation is exacerbated. Any flickering luminous output from a lamp will be reproduced, and very often magnified, in the ceiling, since the ceiling forms a larger area over which the flickering will be detected. Additionally some discharge lamps exhibit a fluctuation in the light output which occurs in sympathy with the cyclic oscillations in the electrical mains supply. This can also be reproduced and magnified in the ceiling.

There are recommended luminance values associated with the optimum lighting conditions in areas illuminated with uplighters. The average luminance of the major reflecting surface, usually the ceiling, should have a value typically between 450 and 550 candela per metre^2 with a corresponding maximum luminance value on the ceiling typically between 1450 and 1500 candela per metre^2.

In many interiors the floor-to-ceiling height of the room will lead to the rejection of the use of uplighters. It is usually accepted that if the floor-to-ceiling height is outside a limiting range (often taken as typically between 2.4 metres and 3.6 metres) then the interior is unlikely to benefit from the use of uplighters. Furthermore the siting of uplighters is critical in respect of adjacent fabrics. Uplighters should not be located in proximity to curtains or blinds since this is likely to create a fire hazard.

There is a clear advantage when using free-standing versions of uplighters in open plan offices where they can be relocated depending upon the layout of the office furniture. Any such relocation of free-standing uplighters must take into account the effects it will have on the uniformity of illuminance it produces in the interior; see Section 11.12.

In interiors where the installation of uplighters is inappropriate then downlighters are used. In such situations, restrictions are placed upon the output characteristics of the luminaires. In an attempt to avoid high luminance reflections appearing on display screens, it is important to use luminaires with an appropriate luminous intensity distribution, which subsequently limits the luminance seen by screen operators.

9.8 Lamp types for use in interiors

The limits placed on office lighting requirements coupled with the luminous and colour output characteristics of lamp types will, when considered in combination, seriously reduce the range of light sources which can be successfully used in interiors.

Filament lamps have good colour rendering properties and in addition they have the obvious benefit of providing a near instantaneous and steady luminous output immediately following switch on. Unfortunately filament lamps are relatively inefficient and coupled with the fact that they produce large amounts of heat during normal operation, they are usually rejected for widespread use in offices. There is an added disadvantage in respect of the generation of heat in that there can be fire risk problems when the lamps are located in proximity to flammable materials.

Discharge lamps, including fluorescent sources, do not attain their steady-state luminous output value immediately following switch-on, but will take time to attain luminous stability. With fluorescent sources this is unlikely to cause problems.[4] However with many other discharge lamps this leads to the rejection of the lamps for use in luminaires used in interiors. Additionally with some discharge lamps, when they are extinguished they will not re-strike immediately if switched on again following a short 'off' period.

The colour characteristics of lamps must be considered when designing office lighting installations. Both colour appearance and colour rendering must be included in a scheme. The colour appearance of some sources will prevent their use in interiors and equally the colour rendering properties of some lamps totally prevents their use in installations where colour discrimination is important. For example, low pressure sodium lamps, by virtue of their monochromatic output, are almost totally unacceptable for use in interiors.

The light output from a fluorescent lamp is strongly dependent on the pressure of the mercury vapour within the discharge tube. It follows that the luminous output will reduce with a corresponding reduction in the value of ambient temperature. Under normal conditions this is unlikely to create a problem in an environment where the ambient temperature is not likely to fall significantly below 20°C.

Compact fluorescent lamps have the benefit that they find use in either downlighters or uplighters. Metal halide lamps are finding an increasing use in uplighters for areas containing display screen equipment. As a consequence of their re-strike capabilities they are limited to use in installations which are switched on and remain on rather than being continually switched. Metal halide lamps have CCT values of typically 3000 Kelvin and a colour rendering index (CRI) value lying between 80 and 90 and are ideal for use in uplighters in areas containing display screen equipment.

9.9 Methods of reducing glare from windows

Several methods are available when striving to reduce the glare introduced into an interior from windows. These can be classified as either temporary or permanent.

9.9.1 Temporary methods

- The use of curtains and blinds, which can be translucent, louvred or slatted, constitutes a temporary method for the prevention of glare.

9.9.2 Permanent methods

- Overhangs above a window will help to reduce the view of the sky from the interior, although there is a simultaneous reduction in the amount of daylight entering the interior.
- The use of low sills will help to increase the illuminance in an interior.
- Increasing the internal fabric reflectances will help alleviate glare.
- Raising the luminance of the window surrounds by strategic siting of the luminaires will avoid excessive luminance contrasts.

- Splaying the window head and sill will assist in reducing the luminance contrast between the window and the window wall.

9.10 Room index

The room index is a numerical value which takes into account the geometry of an interior. It is used in the determination of values of utilization factor, as discussed in Section 9.4, and also in determining the minimum number of measuring points when conducting a lighting survey, as discussed in Section 14.3.1.

The room index (RI) value is calculated from:

$$\text{Room index} = \frac{\text{length} \times \text{width}}{\text{mh (length + width)}} \tag{9.2}$$

where length and width refer to the major dimensions of the room and mh is the mounting height of luminaires above the working plane.

9.11 Automatic control of lighting within offices

The control of lighting within interiors ranges from a simple on/off switching arrangement where one luminaire or a group of luminaires is switched on and off, usually as a consequence of a subjective decision, to a microprocessor-based system incorporating automatic control and associated energy management features.

The simplest form of basic control occurs where switching is achieved manually, dependent almost totally upon a subjective decision at the time of switching. It is of necessity a form of *open loop control*, since there is no automatic feedback. The lamp is switched on and subsequently allowed to attain luminous stability after a period of time. Switching off of lamps under this form of control is often only performed at the end of a working day, whether or not the artificial light is required. It will be evident that this method can be very wasteful of electrical energy.

An improvement on this method is where some form of automatic control is employed which usually incorporates light-sensing photoelectric cells. Such systems are designed to detect levels of prevailing illuminance on the working plane and, using some form of sequence switching, selected luminaires (usually those located adjacent to windows) are switched off when the illuminance on the working plane exceeds some preset value. This method is clearly an improvement on the basic method which relies upon manual switching but there is still room for further advancement. Two possible scenarios can present:

- It is possible that luminaires can be attempting to switch off and on at levels of illuminance hovering around the preset value. As a consequence 'chattering' occurs, whereby lamps will constantly attempt to switch on and off. It follows that some form of 'hysteresis' must be included in the electrical circuitry in order to prevent this from occurring.

• There is likely to be a psychological problem with office occupants who may be severely distracted by lamps which are regularly being automatically switched. This form of annoyance may subsequently lead to visual discomfort experienced by room occupants.

Advances in both lamp technology and electronic circuitry have led to a much more effective method of control, enabling individual fluorescent lamps to be dimmed. Such circuitry, which relies on the production of a high frequency supply to the fluorescent lamp, allows smooth control of the luminous output from its maximum value down to values typically less than 25 per cent of maximum.

It is clear that the ability to control the luminous output from any number of fluorescent lamps in an installation, in sympathy with the level of internal illuminance due to daylight, is a definite advantage over the previously-considered system of switching selected luminaires in their entirety. Such a smooth control of the luminous output from lamps allows the development of a near-constant illuminance level within the office interior, which combines contributions from both natural daylight and artificial light sources.

By definition such control systems are termed *closed loop systems* since they include feedback signals fed to some form of comparator, typically incorporating a microprocessor-based system. Such a system typically includes light-dependent resistors (LDRs) sensing the level of illuminance on the working plane (see Figure 9.6).

The closed-loop control system described has two major advantages in respect of energy saving. It is known that lamps operated on high

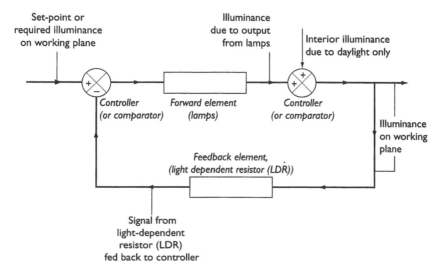

Figure 9.6 Closed loop lighting control system.

frequency supplies exhibit an increase in luminous efficacy, often quoted by manufacturers to be in the order of 10 per cent. Additionally if the demanded luminous output from lamps is controlled in accordance with the internal illuminance due to daylight, then there will be a reduction in energy consumed by the lamps when operating on a reduced luminous output, see Section 7.8.2.

9.12 Emergency lighting

It is essential to provide adequate means of escape from all places of work and public resort in the event of an emergency which necessitates evacuation of personnel. In the UK, emergency lighting is covered by BS 5266[26].
Emergency lighting can take several forms, including:

* emergency escape lighting;
* stand-by lighting;
* escape route lighting;
* open area (anti-panic area) lighting; and
* high risk task area lighting.

9.12.1 Emergency escape lighting

Emergency escape lighting is that part of emergency lighting provided so as to enable occupants of a building to effect a safe means of escape. Such lighting must indicate clearly, and without ambiguity, the escape routes provided so as to allow the safe movement of occupants towards, and out of, emergency exits. Additionally it must ensure that all fire alarm call points, together with fire fighting equipment provided along the escape routes, can be located.

9.12.2 Stand-by lighting

In some interiors it is not possible to effect an immediate evacuation of the building following the occurrence of an emergency or a power failure. Examples of such situations include specified areas within hospitals, e.g. operating theatres, and equipment which relies upon chemical processes where designated 'shut-down' procedures must be carried out. In such situations it is normal to continue with routine activities and stand-by lighting is therefore required. It will be evident that the contribution required from stand-by lighting will be influenced by the routine activities carried out in the interiors together with any associated risks.

9.12.3 Escape route lighting

This usually refers to that part of emergency lighting which is provided so as to enable a safe exit for occupants of a building by the provision of appropriate visual conditions and direction finding on both escape routes and special areas and/or locations. Additionally it is necessary to ensure that all fire fighting and other safety equipment can be readily located.

9.12.4 Open area (anti-panic area) lighting

This usually refers to that part of emergency escape lighting which is provided to reduce the possibility of panic and further to enable safe movement of occupants towards escape routes by providing appropriate visual conditions and direction finding.

9.12.5 High risk area task lighting

This usually refers to that part of emergency lighting provided so as to ensure the safety of occupants of an interior who are involved in a potentially dangerous process or situation and to allow specified shut-down procedures to be completed to ensure the safety of other occupants within the premises.

9.12.6 Luminaires for use in emergency lighting installations

Self-contained luminaires can take on three forms of operation, i.e. maintained, non-maintained and sustained.

In a *maintained luminaire* the lamp is lit continuously. Under normal conditions the lamp is supplied either directly, or indirectly, by the normal supply. When operating under emergency conditions the lamp is powered by its own battery supply.

With a *non-maintained luminaire*, the lamp remains off under normal working conditions and the mains supply to the lamp is also used to charge the luminaire's batteries. Under emergency conditions the lamp is energized from its own battery system.

A *sustained luminaire* is a combination of a maintained and a non-maintained luminaire. A lamp is included in the luminaire which operates from the mains supply during normal working. The luminaire contains a second lamp which, under emergency conditions, is powered from the luminaire's own batteries.

9.13 Sick building syndrome and building related illness

The term sick building syndrome (SBS) is used to describe a host of symptoms which appear to have a high incidence in some buildings and which have a definite work relationship. It is essential to distinguish between sick building syndrome and other illnesses which are connected with buildings or building services. These building related illnesses are typically much less common, they almost always have a clearly identifiable microbiological cause and usually affect relatively few occupants. The term building related illness (BRI) is often preferred for such conditions as humidifier fever and Legionnaire's disease.

Sick building syndrome (SBS) can be caused by a host of factors, one of which is lighting. The symptoms of sick building syndrome (SBS) include:

- headaches;
- nausea;
- dizziness;
- irritation of eyes, nose and throat; and
- lethargy.

Some of problems associated with flicker, which may contribute to the symptoms of sick building syndrome, can be overcome to a large extent with the use of high frequency fluorescent lighting, see Section 7.8.2.

Chapter 10

Lighting for external applications

10.1 Introduction

It will be evident that the criteria for exterior lighting are quite different from those for interior lighting schemes in several ways including:

- marked differences in reflectance values, which subsequently influence values of utilization factor, and hence the efficiency of a lighting installation; and
- the mounting heights of luminaires, which are usually greater for external applications.

The lighting of exteriors for working applications typically demands the illumination of an area for a small number of people who are likely to be involved in work which is of lesser visual difficulty than that experienced in interior applications. The illuminance levels typically required for such external applications are, as a consequence, considerably lower than those required for interior visual tasks.

Examples of exterior lighting installations include factories, floodlighting, security lighting and public lighting.

10.2 Factories

Typical outdoor lighting installations at factories include general external yard lighting, loading bays and storage areas.

10.2.1 General factory external layout lighting

The objective for factory roadway lighting is to provide suitable and sufficient illumination so as to allow the safe passage of personnel both on foot and in vehicles. To this end care must be taken when designing a lighting installation for such areas in order to provide the required illuminance but simultaneously to avoid the development of glare being experienced by works' personnel. It is normal to monitor and control values of average illuminance and point illuminance and in so doing maintain acceptable values of uniformity ratio so as to avoid producing 'patchy' lighting.

10.2.2 Lighting for loading bays

The lighting of loading bays can present problems if attention is not given, at the design stage, to the working patterns of those employed in the area. Loading bays will have vehicles continually arriving at, and leaving from, the working areas and these vehicles may under certain circumstances obstruct the flow of light from luminaires. The resulting shadows formed may constitute hazards for workers involved in visual tasks.

In an attempt to achieve optimum lighting in such areas it can be beneficial to mount luminaires in loading bay soffits, as shown in Figure 10.1, so that the utilization of the available light is high. It is important to avoid producing low values of uniformity ratio which would produce 'patchy' lighting and in so doing render large areas of the loading bay gloomy.

10.2.3 Lighting for storage areas

The principles of lighting open storage areas are similar to those used in the lighting of open spaces. Special examples of storage areas include

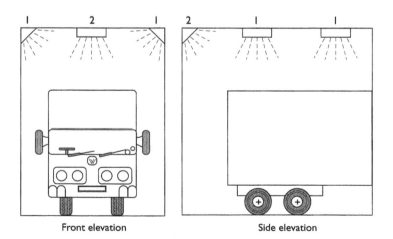

Front elevation　　　　　　　　Side elevation

Figure 10.1 Lighting of loading bays. Location and siting of all luminaires should prevent the creation of glare at all times. 1 Soffit-mounted luminaires will assist vehicle in reversing into loading bay, and side loading. 2 Luminaire directed so as to illuminate the back doors and inside of lorry being loaded.

container yards and storage tanks. Container yards, typically adjacent to railway marshalling yards, usually involve overhead travelling cranes and the lighting design engineer must be aware of the working practices carried out in the area so as not to design an installation which would lead to visual problems. It will be evident that the movement of the crane can obstruct the output from luminaires if they are inappropriately located. This in turn can lead to the production of dangerously deceptive shadows. It is equally important that luminaires do not assist in the development of disability glare in any of the site personnel, including the crane driver.

If the materials being stored are fluids then large tanks are often employed. One of the factors which causes problems with the lighting of such areas is that luminaires on masts or towers are unlikely to illuminate all areas, particularly between adjacent tanks. Figure 10.2 shows a typical layout. In such situations the use of additional luminaires to illuminate shadows will, in many cases, assist in overcoming the problem described.

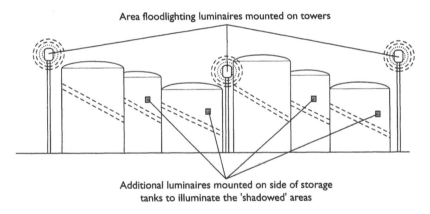

Area floodlighting luminaires mounted on towers

Additional luminaires mounted on side of storage
tanks to illuminate the 'shadowed' areas

Figure 10.2 Lighting of liquid storage tank area.

10.3 Security lighting

The aims and objectives of security lighting are:

- to improve the likelihood of detection, and identification and apprehension of intruders;
- to improve the efficiency of other security measures in use; and
- to improve safety levels for authorized personnel.

Lighting cannot singularly protect premises against the actions of criminals. Lighting is one of a series of components which together can assist in the protection of both personnel and property.

10.3.1 Security lighting for shops and small premises

Owners of shops will be concerned with the security of their premises, especially where the sales rooms and associated areas contain stock which could be of considerable financial value. It will be evident that in such situations the security lighting should pay particular attention to those areas where routine entry to the premises is gained, in addition to those areas of the interior which are visible externally. Security lighting for use in such situations is often influenced by, inter alia:

- Windows and doors that open directly on to the road. In such situations it is unlikely that lighting will be required additional to that provided by public lighting.
- Doors that are set back from the front building line and/or where the contribution from public lighting is considered to be inadequate. In such situations it is often recommended that additional luminaires be provided whose function is to effectively illuminate the doorway. Ideally the additional luminaires should be installed so that they are out of reach of potential offenders.
- Sales areas which are visible externally from the front of the building. Luminaires installed at the rear of the area will highlight the presence of the intruder who will then be visible to members of the public passing by the premises.

When considering security, it is equally important to pay attention to the rear of the premises. Stockrooms are often located at the rear of premises and in such situations it is normal that vehicular access for delivery is also via the rear of the building. If the intention of intruders is to steal quantities of stock from the premises, then they must use some means of transporting the stolen goods away from the rear of the building.

Any lighting provided in the loading area will not only assist with the security of the premises but will also help with the normal operations of authorized personnel within the boundary of the premises.

An additional feature, useful in the fight against crime, is the colour of the loading area. If the area is painted in light colours then the figure or shadow of any unauthorized personnel will be revealed and the conditions should provide a better contrast with the background. It follows that any offender should be more easily detectable.

10.3.2 Security lighting for offices

One of the main considerations when contemplating the security lighting requirement for offices is the size of the premises. With a small suite of offices or a single lock-up office it is difficult to justify the use of dedicated security personnel. Conversely the situation applying with larger office blocks often demands the use of security guards whose sole function is to patrol and guard the offices.

With small office blocks or single lock-up offices, the lighting of the means of exit and entry is of considerable significance. If the premises are situated within their own grounds, then some form of floodlighting installation may prove beneficial. The benefits are twofold in that the floodlighting will aid security and assist in any advertising.

10.3.3 Security lighting for factories

Factories or larger scale premises will create different problems. In addition to the theft of finished goods, raw materials are also targets for criminals and the security of such premises must therefore, of necessity, commence at the factory gatehouse or main entrance.

When security personnel are housed in a well lit gatehouse, typically located at the perimeter of a dark or even unlit site, they are likely to exhibit a reluctance to leave the gatehouse in order to carry out their duties. Additionally, when staff do venture beyond the gatehouse, they are likely to experience adaptation problems when traversing from their well-lit environment to a more dimly-lit one. The duration required for the vision of the security guard to become accustomed to the darkened environment could be sufficient to allow the criminal to escape or alternatively to attack the guard.

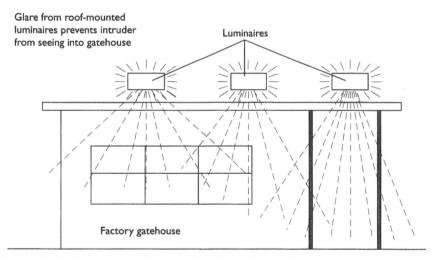

Figure 10.3 Deliberate production of disability glare at factory gatehouse.

Included within the overall security measures available in factories is the deliberate use of disability glare. Large high-intensity lamps installed on the roof of a factory gatehouse can lead to the production of disability glare in a criminal. This would then have the effect of producing indecision in the mind of the unauthorized intruder who would be unsure whether the gatehouse is manned, see Figure 10.3.

The overall illumination conditions required for optimum surveillance of factory premises will be dependent upon, inter alia:

- the luminance of the background;
- the type of surfaces to be illuminated;
- the types of light source used in systems;
- the physical distances at which intruders should be detected; and
- the purpose of the lighting, i.e. for security only or for both security and normal working operation.

When designing a security lighting installation it is essential to avoid the production of shadows. In addition to posing a danger for authorized workers on a site, shadows are likely to provide areas in which criminals are likely to remain undetected and from which they can therefore subsequently make good an escape.

10.4 Floodlighting

Floodlighting can be applied to many installations, e.g. buildings, industrial premises and for sport.

10.4.1 Building floodlighting

Floodlit buildings are commonplace in town and city centres. Thoughtful siting of luminaires and careful selection of the types of light sources used can produce a floodlit effect which is both striking and pleasing to the eye. It is, however, important to appreciate that:

- Floodlighting is not a procedure for illuminating a building during the hours of darkness to the same luminance level, and in the same manner, as that provided by natural daylight.
- The properties and characteristics of the building fabric, with particular reference to surface finish and refelctance values, are highly significant. Figure 10.4 shows the concept of, and effect of, lighting different surface finishes.
- Too much light is as unwanted as too little light when considering floodlighting. The law of diminishing returns applies in that too much light will obliterate shadows and thereby correspondingly reduce the effectiveness of the floodlighting installation.

Building materials such as brick, concrete and stone have non-specular surfaces which are ideal for use in floodlighting schemes. The choice of a suitable light source is greatly influenced by the colour of the building surface which is being illuminated.

For optimum floodlighting of buildings, there should be a flow of light across the front of a building. The direction of this flow should not be identical to that of the direction of normal viewing of the building front. Any contravention of this recommendation will lead to an absence of shadows producing a building appearance that is seen to lack character.

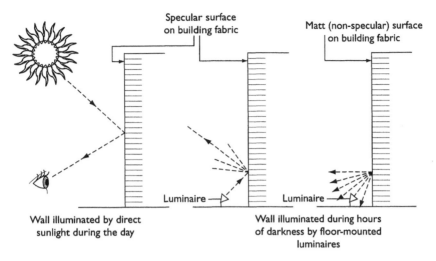

Figure 10.4 Building floodlighting.

It will be evident that the colour output of the light source(s) used in any floodlighting scheme is critical, and the deliberate production of a colour difference can be used to advantage in the process of highlighting different areas of a building.

Table 10. 1 shows typical reflectance values of commonly used building materials.

10.4.2 Industrial floodlighting

The time between dusk and dawn, which is known as *solar time*, approximates to 3780 hours per annum, which is equivalent to approximately 43 per cent of a year. It is clear from this figure that there would be a significant loss in available time if work activities are limited to daylight hours only.

Table 10.1 Typical reflectance values of building materials

Building material	Typical reflectance value (%)
Brickwork (new)	10 to 30
Brickwork (old)	5 to 15
Concrete (new)	40 to 50
Concrete (old)	5 to 15
Plasterboard (new)	70 to 80
Plasterboard (old)	30 to 60
Wood (light varnished oak)	40 to 50
Wood (mahogany and walnut)	15 to 40

Industrial floodlighting is therefore a necessary commodity for those locations where outside work continues during the hours of darkness. Adequate lighting of the correct type will ensure that maximum benefit is obtained from external work activities. Suitable and sufficient working illuminance combined with effective glare control should enable visual task details to be detected clearly, concisely and with speed, which should subsequently allow work to proceed with safety and without creating visual problems for the workforce.

In many industrial activities it is essential to be able to discern the surface colour of engineering and production materials and it will be evident that floodlighting contributes markedly to the ease with which this identification process is carried out. In such situations those light sources with poor colour rendering properties are totally unsuitable and examples of such light sources include low pressure sodium lamps whose output is monochromatic, see Section 7.8.5.

The following points should be taken into consideration when planning the floodlighting of an industrial area:

- Any lighting must be such that it does not present a hazard or cause inconvenience and annoyance to the users of adjacent properties.
- Any lighting installation must be such as not to cause problems and/or be misconstrued as conveying misleading information to users of roads, railways, navigable waterways and aircraft.
- In an attempt to overcome the problem created by shadows, the areas where shadows exist should be 'topped up' using additional luminaires. Such additional luminaires must be strategically sited and carefully aimed so that problems associated with glare are not inadvertently created.

Techniques used in industrial floodlighting often incorporate either wide-angle luminaires, typically mounted in proximity to the area to be lit, or alternatively high-mast lighting. It is possible in some situations that a combination of both methods will provide an effective solution.

The use of wide-angle luminaires for area lighting will in general produce greater illuminance values on the vertical faces of stacks of materials whilst high-mast lighting is often used when lighting large surface vehicle parks. If an area contains goods which are highly-stacked then the mounting height of luminaires used in the flood-lighting installation must be sufficiently high so as to allow penetration of the light output into the rows and walkways between the stacked material.

Metal halide lamps along with high pressure sodium lamps tend to be used often in floodlighting schemes whilst tungsten halogen lamps are used frequently when 'topping up' shadowed areas, which, if left unlit, might cause potentially dangerous problems.

Automatic control of floodlighting installations is usually employed. This invariably involves the use of photoelectric cells which are fail-safe, i.e. should there be a fault with the photoelectric cell the lamps will remain switched on, thereby alerting maintenance personnel of the malfunction in the system.

10.4.3 Sports floodlighting

There is a clear demand for leisure and sports facilities, both indoor and outdoor. This demand is due in part to the increased awareness of the benefits in relation to the general fitness and quality of life that sport and exercise brings. It will be evident that in order to obtain full potential from such facilities, some form of floodlighting should be installed wherever possible.

The illuminance provided for outdoor courts and areas at leisure centres can be substantially lower than that required for senior level and professional sport. In such situations colour television transmission requirements (in particular illuminance values and correlated colour temperature (CCT) values) have to be satisfied.

For installations at leisure centres, luminaires mounted on columns are frequently used. At professional sporting venues, floodlighting installations typically involve:

- luminaires on floodlighting towers (typically located in the corners of the floodlit area);
- individual luminaires mounted typically on grandstand fasciae, i.e. an absence of towers and/or poles;
- a combination of both of the above techniques.

External lighting, including floodlighting and public lighting, is often achieved using light sources where the output from luminaires travels through the atmosphere before eventually reaching its working plane. During this journey through the atmosphere the light will encounter moisture and general air-borne particles, the effect of which is to reduce the light energy arriving at the working plane. The losses, referred to as *atmospheric losses*, are influenced by:

- time of day;
- mounting height of the luminaires;
- season of the year;
- geographic location.

Under certain extreme conditions atmospheric losses can be up to 25 per cent.

Further information in respect of lighting for sport can be found in CIBSE LG4: Sport.[27]

10.5 Public lighting

Public lighting can be defined as any lighting provided for the public use which is usually maintained at the public's expense. The functions of public lighting can be classified as:

- to ensure the continued safety of road users and pedestrians;
- to assist the police in the enforcement of the law;

- to improve the environment for the benefit of residents; and
- to highlight shopping areas and areas of civic importance.

10.5.1 Road lighting

When travelling on a road in a vehicle, a driver will normally take in the whole scene but equally he or she will only give conscious attention to a limited amount of the information presented to him or her by the scene. Experience gained will teach a driver, by using a form of sub-conscious selection, to focus attention on any particular hazard and in so doing act accordingly.

When a driver travels along a road during the hours of darkness, objects are seen by one of two processes, either (a) direct vision (using surface detail) or (b) by *silhouette vision*. A driver will see as silhouettes either small objects at medium distances or large objects at greater distances. By directing the beams from the vehicle headlights on to the road so that they strike it at glancing angles (towards oncoming motorists) the whole of a road surface can be made to appear bright.

Careful aiming of lighting in a downward direction is required so that drivers can see nearby objects by surface detail. It is vital that vehicle drivers be able to see road characteristics, e.g. road markings, kerb edges, and road surface texture so that it is possible to formulate a complete picture of the scene ahead. When a motorist is in possession of this information he or she can judge, with an acceptable level of accuracy, the position of their vehicle relative to the road.

Other hazards likely to cause problems for a vehicle driver include:

- glare from the headlights of oncoming vehicles; and
- other lights on other vehicles which appear excessively bright; these include stop lights, direction indicators, reversing lights and fog lights.

For optimum road lighting conditions, the designer should strive for the following:

- a uniform road surface luminance;
- adequate and acceptable illuminance;
- suitable silhouette contrasts of the road ahead; and
- glare control from road lighting luminaires.

Low pressure sodium lamps have the highest luminous efficacy of artificial light sources and have been used extensively for road lighting. Their colour rendering properties are very poor and many public lighting engineers use such monochromatic sources only for minor road lighting installations. High pressure sodium lighting is now used on many trunk roads, other major roads and in town centres and areas of civic importance. It is important to appreciate that when lighting highways in the vicinity of railway lines, the use of low pressure sodium lamps could cause confusion to train drivers who are looking for coloured signal lamps for the control of rail traffic, as shown in Figure 10.5.

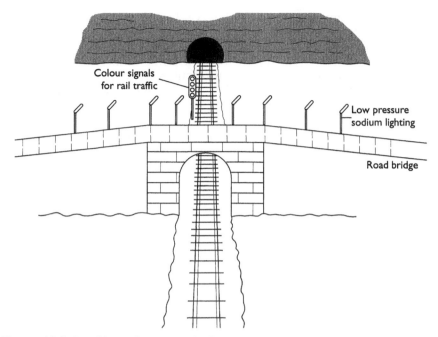

Figure 10.5 Possible conflict caused by low pressure sodium road lighting in proximity to colour signals used for rail traffic control.

Many level crossings are fitted with some form of closed circuit television (CCTV) monitoring which transmits pictures to a control centre. There is a special requirement in relation to the television cameras in that they will only operate efficiently, and thereby produce clear and unambiguous pictures, when the light sources and television cameras are matched in relation to colour characteristics, all as detailed in Section 10.7.

10.5.2 Tunnel lighting

The purpose of lighting in road tunnels is to allow traffic to continue to flow through the tunnel at similar speeds, and with the same degree of safety, as that applying on the tunnel approaches. One significant difference between lighting for open roads and lighting for tunnels is that in tunnels lighting is required throughout the whole day. Under open road conditions the driver of a vehicle should be able to see a specified distance ahead. As the vehicle approaches a tunnel, this distance will project into the tunnel entrance and there should therefore be lighting installed within the tunnel in order to ensure that the visibility level of the driver will be maintained.

The eyes of the driver, during the approach to and entry into the tunnel, will gradually become accustomed to the darker surroundings. This adaptation is a continuous process and so therefore as the driver proceeds even further into the tunnel, the road surface illuminance can be gradually reduced if the tunnel is sufficiently long. In addition to striving for optimum forward driving conditions, it is equally important to maintain rearward visibility when leaving a tunnel so that a driver can continue without experiencing major visual problems.

The location of individual luminaires within a tunnel is highly significant. If they are installed in discontinuous rows, either along the roof or on the wall side, it is possible that a flicker effect will be developed, which can produce serious problems for the vision of vehicle drivers. Furthermore images of the luminaires, if so mounted, can also be seen in the bonnet of a vehicle which can produce the same effect. Additionally discontinuous rows of luminaires can give rise to a low uniformity ratio value in respect of tunnel illuminance.

Typically continuous rows of fluorescent luminaires are used for interior zones and for optimum effects they should be located axially along the tunnel length. For lighting of entrances and exits, sodium lamps are often used.

Road lighting is included in BS 5489.[28]

10.6 Lighting for externally located advertising signs

When considering the external illumination of informative and advertising signs, several factors must be taken into account, including:

- the creation of a scenario where the text on the sign is clear, unambiguous and free from reflected glare; and
- the prevention of stray reflections or direct glare (from the luminaires used or from the sign itself) from becoming an environmental nuisance to others in the proximity.

If signs are located such that they are viewed from below, as shown in Figure 10.6, then it is usual to provide luminaires mounted on brackets and affixed so that their output is directed onto the sign from above. If, however, signs are located such that they are viewed from above, or mounted on the ground, as shown in Figure 10.7, then luminaires are often installed below the sign with their light output directed upwards.

In all cases it is essential that the mounting and aiming of luminaires used is such that stray effects are minimized. It is equally essential that the lighting installation conforms to all relevant legislation, including any local authority restrictions. Failure to comply with current regulations will, in addition to rendering those responsible liable for the consequences, also result in the installation contributing to light pollution, as discussed in Section 10.8.

10.7 Lighting for use with closed circuit television (CCTV) systems

Many lighting schemes are employed in areas where closed circuit television (CCTV) systems are installed, e.g. security lighting, floodlighting

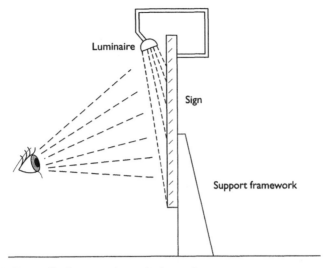

Figure 10.6 Externally-illuminated sign lit from above.

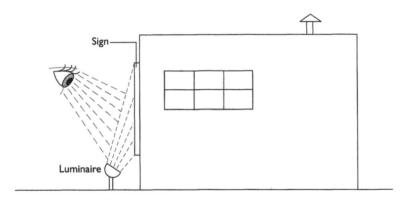

Figure 10.7 Externally-illuminated sign lit from below.

and public lighting. It is important to appreciate that television cameras used in such systems usually have specifications which require compatibility with the visual scene e.g. in respect of correlated colour temperatures and spectral sensitivity. The correlated colour temperature of the camera must be matched with that of the scene being filmed and this limitation will therefore preclude the use of some types of light source, e.g. low pressure sodium. The spectral sensitivities of television cameras used in CCTV schemes are unlikely to be the same as that of the human visual system. The selection of an appropriate light source for use in such areas must therefore also take into account the spectral sensitivity properties of the CCTV cameras used.

10.8 Light pollution

There seems little doubt that extraneous light, in various forms, is a public enemy of increasing proportions. Astronomers are often particularly aggrieved in respect of this offensive light insomuch as they would prefer to view the night sky with as little stray light as possible. The situation, which is particularly annoying for astronomers, occurs for a variety of reasons but in general terms it is due to either badly designed lighting installations or inaccurately aimed luminaires or a combination of both.

The likely consequences are sky glow or light pollution, where artificial lighting spills over, and therefore trespasses into, areas for which it is not intended. The situation becomes more pronounced when light is scattered through the night sky by particles of dust and droplets of water.

There appears little doubt that light pollution is a serious problem and unless the situation is addressed and subsequently reversed there is every possibility that the view of the night sky and all of the details contained therein, will be lost. Light pollution is considered by many to be as obtrusive and offensive as pollution caused by other agents, e.g. noise, smoke and chemicals.

10.8.1 Causes

One of the factors leading to the development of light pollution is the effects of some public lighting, although arguably other lighting installations, e.g. floodlighting and security lighting, can create more severe effects. Furthermore, the colour of the light output from lamps is of paramount importance.

Both high pressure and low pressure sodium lamps have been used extensively for public lighting but their colour appearance and colour rendering properties need to be considered. Low pressure sodium lamps emit monochromatic light which produces very poor colour rendering. It will be evident that if light is trespassing onto territories which are not intended to be lit, then electrical energy will be wasted. It is clearly in everybody's interest to minimize stray light, and in so doing reduce electricity consumption which will produce a corresponding saving in electrical energy charges. Furthermore the earth's finite energy resources will be conserved.

Since the introduction of the filament lamp, there have been amazing advances in lamp technology which have led to towns and cities being illuminated. If the towns and cities are lit, with due consideration given to the likely effects of extraneous light, the detail in the night sky will once again be visible. To this end local authorities can play a major role in developing, implementing and monitoring strategies for lighting schemes, both road lighting (which is usually their own responsibility) and floodlighting (both industrial and sports) which they can control through planning regulations.

10.8.2 Remedies

Some methods of reducing stray and obtrusive light are simple; these include:

- The development and implementation of a switching strategy, i.e. switching off of external lighting when not required. In addition to the reduction in light pollution there are the attendant energy savings.
- Resisting the temptation to 'overlight' areas. Some areas which are 'washed' with light can still be lit to an acceptable level with either a reduction in the number of lamps used or a reduction in the rating of lamps, or both.
- Striving for increased attention to be given to the aiming of light output from luminaires. As a general rule it is beneficial to aim luminaires in a downward direction rather than in an upward one. One possible offender in this respect is the lighting of directional signs on motorways. Often, and typically due to the lack of available mains electrical supplies alongside the hard shoulder, motorway signs are lit using lamps supplied from portable sources. The difficulties encountered in mounting luminaires so that they overhang the front face of the sign usually means that they are mounted underneath the sign, with the light therefore being directed in an upward direction. This contributes to light pollution. If it is not possible to aim luminaires downwards, then shields and/or baffles can be used to advantage.
- The minimizing of glare. Any glare sources, in addition to contributing to light pollution, will be distracting to drivers and pedestrians and may, in certain circumstances, be a contributory factor in road traffic accidents.

The aims and objectives of any lighting strategy should include the following points:

- the complete integration of light sources within a residential, commercial and/or industrial complex in order to achieve a harmonious combination;
- an increase in the quality of area lighting which will subsequently reduce the quantity of light required;
- careful consideration when floodlighting premises and buildings of civic importance;
- an attempt to strive for a balance between lit areas and surfaces, and those deliberately requiring to be left with shadows so as to obtain a suitable reduction in unnecessary 'overlighting' and 'underlighting';
- due consideration to lighting used for advertising, bad practice of which is a major contributor to light pollution;
- due consideration to the quantity and directional characteristics of stray lighting from buildings, which could reduce the amount of light specifically required for advertising; and
- restriction of the light sources used to those with acceptable colour characteristics.

Chapter 11

Visual task lighting

11.1 Introduction

Many everyday visual activities involve reading, writing, etc., primarily encompassing work in two dimensions only. Similarly, other working practices will, of necessity, demand good sight and vision in three dimensions. Good dimensional acuity is influenced significantly by the level of stereoscopic vision. Workers who show an absence of stereoscopic vision will be seriously disadvantaged and in some cases may be debarred from carrying out certain occupations. In the absence of suitable and sufficient lighting, the benefits of good stereoscopic vision will not be appreciated.

If a target is lit using a small source which is uni-directionally incident on the target, it is highly likely that shadows will be produced which contribute to the development of an appearance of solidity. Such a scenario will fail to correctly reveal form and surface texture of objects.

For optimum lighting conditions it is highly desirable to produce a scheme which combines both direct light and diffuse light. This will assist in the illumination of shadows, which in turn prevents them from being a potential source of accidents. Natural daylight has a combination of direct sunlight and diffuse skylight; this subsequently produces modelling which allows the immediate appreciation and recognition of the form of objects.

In an attempt to reduce the likelihood of occupational accidents, it is essential to consider several factors. It is totally unacceptable to design a system singularly on the basis of the minimum level of illuminance required. When considering a task lighting installation, it is important to consider, for example, luminance distributions and stray light effects.

11.2 Importance of lighting in the workplace

The greater proportion of information received from the world around us is visual information. Vision can be considered to be the most important of man's senses since without light man cannot see. The main purpose of lighting is to enable individuals to see. It is often thought that the importance of good lighting is rarely appreciated and support for this hypothesis is given by the fact that:

- lighting is often taken for granted and it is rarely questioned as to whether an installation could be improved;
- the operation of the human visual system, including the link with the brain, can appear to compensate for poor lighting and in many normal visual activities, individuals are rarely aware of some of the problems attributed to poor lighting.

Seeing is based upon a minimum of obtained visual information and relies heavily upon memory of previous visual experiences. It will be evident therefore that if lighting in a particular workspace is poor, workers may not readily appreciate the situation because the images ultimately produced are even more heavily dependent upon interpretation with a correspondingly reduced contribution from 'real' information.

It is generally accepted that poor lighting means that the human visual system will not operate at its optimum efficiency. There will be associated ramifications if an individual is subjected to poor lighting conditions including:

- 'real' information will be lost and a worker is therefore considered more likely to make errors; and
- poor visual signals result in information taking longer to produce an image, which can cause problems if a worker is involved with moving objects. It will be evident therefore that good task lighting is essential in the workplace.

11.3 Analysis of the working visual task

A comprehensive analysis of the visual task presented at the workplace is essential when designing a lighting scheme for a particular working practice. Figure 11.1 shows the salient components of an occupational visual task analysis.

The health and safety practitioner must appreciate all of the factors which influence the ability to see detail involved in a workplace task so that optimum task lighting conditions will be produced. These factors can be categorized into:

- the characteristics of the individual observer; and
- the characteristics of the visual task itself.

Before embarking on a full survey of the working practices it is important to address the following questions:

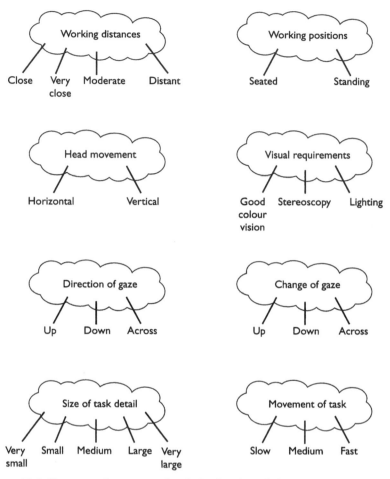

Figure 11.1 Elements of an occupational visual task analysis.

- Are the details of the task easy to see under normal working conditions?
- Is the visual task likely to be undertaken for excessive time periods?
- If there is inadequate task lighting and as a result errors occur are the consequences considered to be serious?

Once these questions have been answered the next stage in the design is to consider the expectations of task lighting. This can be achieved by considering:

- Those features which task lighting should highlight. These include physical size, surface qualities, surface colour.
- Those features which task lighting should help prevent. These include deceptive shadows and glare.

Those characteristics of the observer's performance which clearly influence his or her ability to see detail include:

- the individual's sensitivity to physical parameters;
- transient adaptation time taken when an individual changes scenes of differing luminance;
- the exposure time, the length of time for which the task is undertaken;
- the age and general state of well-being of an individual observer.

Those characteristics of the visible task itself which will ultimately influence an individual observer's ability to see detail include:

- the detail encompassed within the task;
- the luminance contrast of both the task and the background against which the task is viewed; and
- the specularity of the detail contained within the task.

11.4 Visibility of an object

Different forms of energy are utilized by the human body: one of these, light energy, is frequently regarded as being the most important. Light is essential in allowing objects to be seen and it is equally vital in order to be able to distinguish certain characteristics of objects, e.g. form, colour, seen in our daily activities.

Such characteristics as state of mind and level of fatigue are affected by both the lighting conditions applying and also by the colour of objects seen within an individual's field of view.

The luminance of the following will significantly affect the ease with which the human body detects an object:

- the working task itself;
- the immediate surround; and
- the distant surround.

It is well documented that the eye perceives variations in the luminance of objects and their surroundings or alternatively between different areas of the same object.

Workplace accidents are often blamed on inadequate lighting conditions. The errors occurring as a consequence are often attributed to individuals who have difficulty in correctly identifying objects. The eye is capable of adapting to scenes where the illuminance and hence luminance levels may be low and so often these deficiencies or malfunctions within a lighting installation may go unnoticed.

11.5 Factors influencing the ability to see detail

The amount of light required for an accurate interpretation of a given visual task is significantly influenced by four factors, which are interdependent:

- the physical size of the target object;
- the luminance contrast between the object and its surroundings;
- the reflectance properties of the surroundings; and
- the time during which the target object is viewed.

Some degree of control can be exercised over three of these factors, the exception being that of the size of the target object. Typical reflectance values of common workplace materials are shown in Table 11.1.

Table 11.1 Typical reflectance values of common workplace materials

Material	Approximate reflectance value (%)
Medium quality white paper	75
Aluminium	75
White cloth	65
Bright steel	25
Cast iron	25
Matt black paper	05

11.6 Quality of illumination

Much of the treatment of task lighting is undertaken on the assumption that the incident light provided is 'white light', i.e. the prevailing light contains reasonable quantities of light over the entire range of the visible spectrum. It is well-documented that for example, a red material will reflect red wavelengths of light incident upon it and will thus absorb all other wavelengths. It follows therefore that in addition to striving for optimum values of both illuminance and luminance, the quality of illumination must be taken into account when considering visual efficiency.

Coloured light can be used to good advantage in some situations with a possible enhancement of colour contrasts which may, in the absence of coloured light, be low. In general terms, if it is required to increase the contrast between two coloured surfaces, then light which is deficient in the wavelengths of the darker of the two surfaces may be used.

11.7 Visual comfort

When considering the health and safety of individuals in the workplace, visual capacity and visual comfort are highly significant. Workplace accidents are often due, inter alia, to inadequate lighting conditions and the resulting errors made by the individuals who consequently have difficulty in correctly identifying objects.

Deficiencies in lighting conditions which ultimately lead to visual disorders are not uncommon in the workplace, although the ability of the eye to adapt to such situations sometimes masks the underlying problems.

Some of the factors which will assist in the production of optimum visual comfort include:

- constant illuminance levels;
- constant luminance levels;
- glare-free environment;
- optimum luminance contrast levels;
- acceptable colour contrast;
- reduction in visual distractions; and
- elimination of flicker and/or the stroboscopic effect.

When analysing the visual task it is important to consider both the workstation in general and the detail required on the task.

The luminance of an object, of its immediate surroundings and of the work area in general will significantly affect the ease with which it is detected. What the eye really perceives are variations in luminance of objects and its surroundings or conversely between different parts of the same object. Figure 11.2 shows methods of shielding unwanted distractions from adjacent working areas.

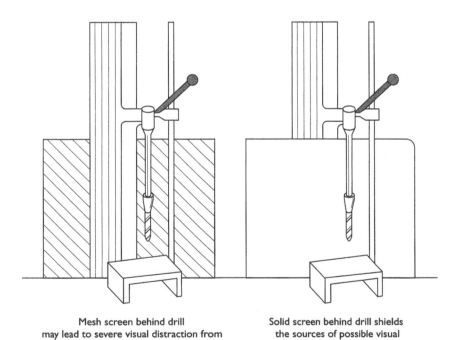

Mesh screen behind drill
may lead to severe visual distraction from
adjacent workers or from static fixtures

Solid screen behind drill shields
the sources of possible visual
distraction from view of drill operator

Figure 11.2 Shielding of unwanted visual distractions.

11.8 Glare in the working environment

The visibility of a working task is influenced to a large extent by any sources of glare within the visual field. One definition of glare is that it is any excessive variation in luminance within the visual field. Glare can be conveniently divided into two main groups, i.e. *disability glare* and *discomfort glare*.

Glare can be thought of as 'direct' when it occurs as a consequence of bright sources directly in the line of vision, or alternatively as 'reflected' when light is reflected onto surfaces which have high reflectance values. Factors involved in the production of direct glare include the luminance of the light source and the location of the light source.

A form of glare which will disable an individual from carrying out a particular visual task is referred to as disability glare. An everyday example of disability glare occurs when an individual looks at the headlights of a stationary vehicle during darkness. Under these circumstances it is impossible to discern the scene at the sides of the vehicle immediately behind the headlights. The glare disables the individual from carrying out the visual task, see Figure 11.3.

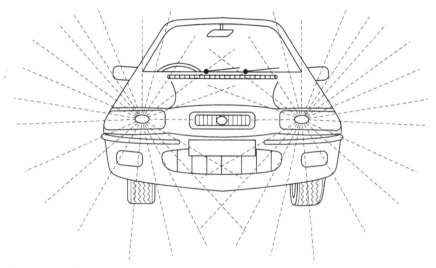

Figure 11.3 Effects of disability glare.

The magnitude of the disabling effect experienced with disability glare is unlikely to occur with discomfort glare. An individual will experience a feeling of discomfort when the exposure time is prolonged.

11.8.1 Control of discomfort glare

Disability glare and discomfort glare are the major forms of glare and are considered in detail in Sections 5.2.1 and 5.2.2. Discomfort glare is more prevalent in interiors and often occurs as a consequence of either a badly-designed lighting installation or a change of use of the interior from that which it was originally intended. In addition, it is possible to assign numerical values to the degree of discomfort glare prevailing in a given interior. It is therefore possible, at the design stage, to eliminate any such adverse effects.

A method used for calculating the 'limiting glare index' (LGI) is shown in detail in the CIBSE (Chartered Institution of Building Services Engineers) Code for Interior Lighting 1994.[29] The method is a 'step-by-step' process where an initial and uncorrected glare index value is obtained from published tables. In these tables the major dimensions of the room interior (length and width) are given in multiples of the mounting height H which is taken as the distance between the horizontal lines passing through the eye level of a seated observer and the centre line through the luminaire(s). The eye level of a seated observer is taken, initially, as 1.2 metres above floor level.

A second stage to the calculation involves applying two correction factors to the initial glare index value obtained. These factors involve variations in (a) the luminous flux of the luminaire and (b) any variation in mounting height from the normal seated eye level of 1.2 metres.

Once the 'final glare index' value is calculated, it is compared with reference values of limiting glare index (LGI), which are published in references, e.g. CIBSE Code for Interior Lighting 1994.[29] Should the final glare index value calculated be lower than the LGI value, the development of discomfort glare is unlikely.

11.8.2 Direct and reflected glare

It is also possible to categorize glare as either 'direct' or 'reflected'. Direct glare occurs when the origin is bright sources. Reflected glare occurs when light is reflected from specular or mirror-like surfaces. When considering glare it is essential to consider the following:

- the luminance of the source;
- luminance distribution;
- the position of the source; and
- the time duration of exposure.

The maximum value of luminance which is tolerable by the eye as a result of direct viewing is approximately 7500 cd·m^{-2}. Glare occurs when the light source falls within a 45° angle of the line of sight of an individual, as shown in Figures 11.4 and 11.5. Figure 11.6 shows an example of reflected glare where a specular surface reflects direct sunlight.

The visual comfort of an individual is influenced by the distribution of luminance across the immediate field of vision and furthermore by the luminance of the environment when seen by individuals who glance away from their work.

Ideally the work surrounds should be less bright than the work itself. Optimum visual comfort occurs when the work is slightly brighter than

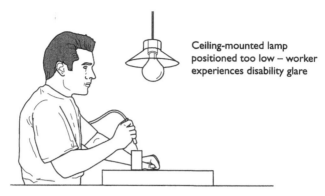

Ceiling-mounted lamp
positioned too low – worker
experiences disability glare

Figure 11.4 Low suspended lamp incorrectly positioned.

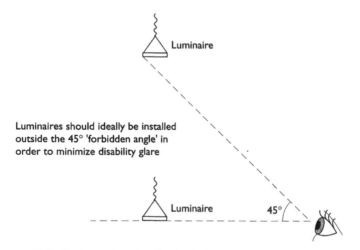

Luminaire

Luminaires should ideally be installed
outside the 45° 'forbidden angle' in
order to minimize disability glare

Luminaire   45°

Figure 11.5 Optimum location for luminaires.

the near surround. This in turn should be slightly brighter than the far surround.

The ideal ratio for distribution of luminance across a task is usually taken as 10:3:1. It is possible to obtain a 'trade-off' in luminance values, away from a visual target by using different materials. Figure 11.7 shows the ideal luminance distribution across a visual task.

Whilst glare is normally associated with bright light sources, those sources with a relatively low luminance value can still produce glare if the exposure duration is prolonged.

One of the main sources of glare in an interior is caused by direct sunlight which is admitted via windows. Whilst the well-being of occupants is important, and daylight should be admitted wherever possible, the control of glare from windows can cause problems in the workplace. Methods of reducing glare from windows are given in Figure 11.8.

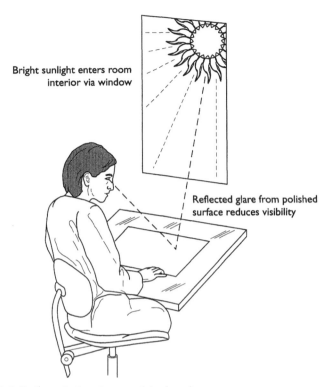

Bright sunlight enters room interior via window

Reflected glare from polished surface reduces visibility

Figure 11.6 Reflected glare from polished surface.

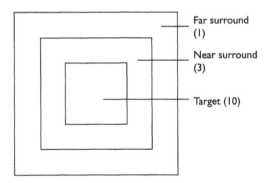

Far surround (1)

Near surround (3)

Target (10)

Figure 11.7 Ideal luminance distribution across a visual task.

11.9 Veiling reflections

The luminance contrast of a visual task will be influenced by, inter alia, the reflectance values of the task and furthermore upon the manner in which the task is lit. Task materials with matt finishes will reflect the

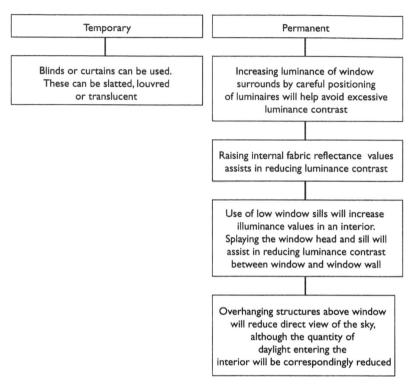

Temporary	Permanent
Blinds or curtains can be used. These can be slatted, louvred or translucent	Increasing luminance of window surrounds by careful positioning of luminaires will help avoid excessive luminance contrast

Raising internal fabric reflectance values assists in reducing luminance contrast

Use of low window sills will increase illuminance values in an interior. Splaying the window head and sill will assist in reducing luminance contrast between window and window wall

Overhanging structures above window will reduce direct view of the sky, although the quantity of daylight entering the interior will be correspondingly reduced

Figure 11.8 Methods of reducing glare from windows.

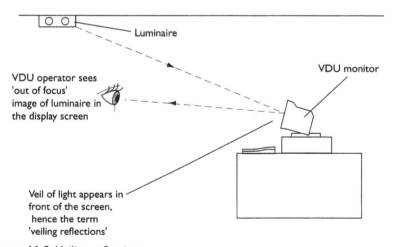

Luminaire

VDU monitor

VDU operator sees 'out of focus' image of luminaire in the display screen

Veil of light appears in front of the screen, hence the term 'veiling reflections'

Figure 11.9 Veiling reflections.

incident light equally in all directions and it follows that the direction of the incident light falling on the task material is unimportant. Conversely if the task material has a specular finish then the direction of the incident light is important.

When the image of a source of high luminance, e.g. a luminaire or the sun, is reflected from a surface being viewed by an individual then veiling reflections will be created. An example of this is experienced when looking at a display screen. If the geometry of the individual, screen and luminaire is not controlled, an out-of-focus image of the luminaire will be seen in the screen. This will throw a veil of light on the front of the screen, hence the term veiling reflections, and some of the text on the screen will become illegible (see Figure 11.9).

Downlighters with a strong downward component of light distribution will produce a significant loss of task visibility due to veiling reflections. A lighting installation should ideally provide a minimum directional component immediately above the task itself. Light should preferably reach the task from wider angles which will reduce glossy reflections seen by the individual in the visual task. In practice therefore luminaires with a *batwing* light distribution, as shown in Figure 11.10, will ensure that light is directed onto a visual target from wider angles, thereby reducing the likelihood of producing veiling reflections.

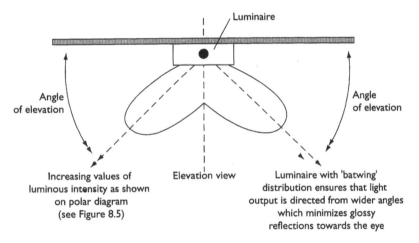

Figure 11.10 Typical luminous intensity distribution from 'batwing' luminaire.

11.10 Use of daylight for task lighting

In the UK, it is generally accepted that daylight is singularly unlikely to be able to provide sufficient internal illuminance to enable visual tasks to be carried out at all times. It follows that lighting calculations for room interiors rarely take into account any contribution from natural light since it is such a variable quantity. Clearly during late afternoons in December (in the UK) the illuminance due to daylight both externally and internally is negligible. Despite this obvious disadvantage, daylight must be harnessed and use made of it whenever possible. Daylight within an interior is free, if the building capital costs associated with the provisions of windows are neglected. Furthermore it is highly desirable that a link is maintained between the exterior and the occupants of an interior. If this link is absent, individuals within the interior would quickly lose the feeling of 'well-being'. Often the repercussions of this situation are a reduction in worker output productivity.

Additionally it is important to appreciate that natural light has several beneficial effects on the human body, including:

- it affects body rhythms, for example sleep patterns and hormone secretion; and
- it influences physiological performance, e.g. moods and levels of alertness.

It is well documented that whilst natural light should be harnessed to its full potential, there is a practical limit to the quantity admitted into interiors, although this is influenced by non-lighting constraints. Too many windows will allow corresponding excesses of heat transfer between the interior and exterior of a building during both summer and winter.

11.11 Luminance contrast and contrast rendering factor (CRF)

The design of lighting strives to achieve a uniform illuminance distribution across a working plane. Such a scenario may, however, produce poor lighting of the visual task itself. Bright reflections in a visual task will reduce its visibility and they are likely to give rise to discomfort. In the process of seeing, the human eye discerns details by discriminating between the lighter and darker parts of an object or a visual task. The changes in luminance (between the lighter and darker parts of the task) can be assigned a numerical value. This value is referred to as the *luminance contrast*, and, by definition, it is calculated from:

$$\text{Luminance contrast (C)} = \frac{L_t - L_b}{L_b} \tag{11.1}$$

where:

L_t = Luminance of target $(\text{cd} \cdot \text{m}^{-2})$
L_b = Luminance of background $(\text{cd} \cdot \text{m}^{-2})$

It is also possible to determine values of luminance contrast from the corresponding reflectance values of both the target and the background (providing the illumination remains constant) from the expression:

$$\text{Luminance contrast (C)} = \frac{R_t - R_b}{R_b} \tag{11.2}$$

where:

R_t = Reflectance of the task
R_b = Reflectance of background

It will be evident that visual tasks contain materials each of which may have a different specularity. It follows therefore that contrasts will be ever-changing as the light falling on the task changes. It will be apparent that the values of task contrast give insufficient and inadequate information about the true effects of the lighting conditions.

It is well documented that the eye will tend to be attracted towards the brightest part of the visual field. This phenomenon is referred to as *phototropism*. The phototropic effect is used to advantage in display and advertising.

Consider the routine visual task of reading black print on white paper. If the white paper is placed on a black-topped desk and then illuminated with a desk lamp only, the remainder of the room will appear in darkness. Due to the phototropic effect, the eyes will be drawn to the white paper and so therefore maximum concentration on the task will be achieved. However this maximum concentration is achieved at the sacrifice of some visual comfort. The extent of the visual discomfort experienced will depend upon the illuminance on the paper.

If, however, the paperwork is placed on a pale coloured desk, e.g. cream, and the room is illuminated with general lighting, the conditions developed will produce maximum visual comfort but at the expense of reduced visual attention. It follows that, when attempting to obtain optimum conditions, a balance must be made between the need to avoid visual discomfort and the simultaneous need to prevent visual distractions.

It is possible to overcome the apparent ambiguity caused by the constantly changing parameters previously described. If the luminance contrast is measured under the lighting conditions of interest and then compared with the luminance contrast value obtained using reference conditions, it is possible to calculate a parameter referred to as *contrast rendering factor* (CRF).

CRF is defined as:

$$\text{CRF} = \frac{C_1}{C_2} \tag{11.3}$$

where:

C_1 = luminance contrast of target under lighting conditions of interest
C_2 = luminance contrast of target under reference lighting (viewed from the same direction)

In general, the value of CRF is specific to a particular target, at a particular point, under a particular lighting installation and viewed from a particular direction.

The value of CRF obtained in the manner described previously is a relative indication of the ability of a particular lighting installation to produce a combination of high contrast together with good visibility. In general, values of CRF lower than 0.7 are likely to lead to lighting conditions which will be unacceptable.

11.12 Uniformity ratio and diversity ratio

When designing a lighting scheme it is almost inevitable that the illuminance over the whole of the working plane will not be constant. An indicator of the 'constancy' of illuminance over a task area and its immediate surround is given by a parameter referred to as the uniformity ratio.

$$\text{Uniformity ratio} = \frac{\text{Minimum illuminance}}{\text{Mean illuminance}} \qquad (11.4)$$

Lighting design must be considered carefully and it is important not to produce a system which will ultimately create large variations in illuminance across an interior. A parameter referred to as the diversity ratio is used to indicate the variation, where:

$$\text{Diversity ratio} = \frac{\text{Maximum illuminance}}{\text{Minimum illuminance}} \qquad (11.5)$$

11.13 Spacing-to-height ratio (SHR)

The spacing-to-height ratio (SHR) is the ratio of the spacing between the centre of adjacent luminaires, in an installation, to the height between the horizontal centre line of the luminaires and the working plane. Consider Figure 11.11, which shows, in elevation, an office interior.

Two values are often used, i.e.

- SHR NOM – nominal spacing-to-height ratio; and
- SHR MAX – maximum spacing-to-height ratio.

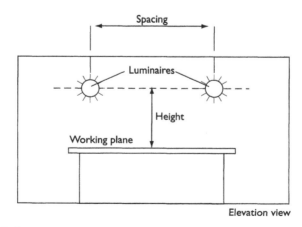

Figure 11.11 Spacing-to-height ratio.

It will be evident that the SHR, which is dependent upon the luminous output distribution of the luminaires used, will influence the uniformity of illuminance on the working plane.

11.14 Influence of illuminance on human visual characteristics

Visual acuity and visual performance (the ability to carry out a particular visual task with both speed and accuracy) are both influenced by prevailing illuminance. In both cases it is apparent that the relationship follows the law of diminishing returns. Figure 4.17 shows the relationship between visual acuity and illuminance and Figure 11.12 shows the relationship between visual performance and illuminance for differing levels of size and contrast.

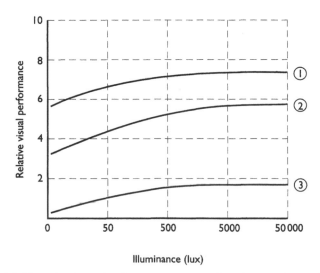

Figure 11.12 Typical relationship between visual performance and prevailing illuminance. 1 Large size, good contrast; 2 Medium size, medium contrast; 3 Small size, low contrast.

11.15 Flicker and the stroboscopic effect

When the frequency of flashing lights increases from a relatively low value, the appearance of a flickering light will be produced. If the rate of flashing is increased still further, the sensation of a flickering light disappears and the light output appears as a continuous source.

Critical fusion frequency (CFF) is defined as the frequency at which the perception of a flickering light disappears and the sensation of a continuous light begins.

Flicker is an obvious source of discomfort and distraction in individuals and in some cases it can trigger an epileptic seizure. Individual sensitivity to flicker varies markedly. Human perception of flicker is influenced by the frequency of oscillations, the amplitude of modulation and the area over which the modulation occurs. The combination of large variations in amplitude occurring at low frequencies and over large surface areas will produce the most noticeable light modulation.

Advances in lamp technology combined with the introduction of high frequency operation have helped to almost totally eliminate the problem of flicker from some fluorescent sources.

If the rate of flicker reaches a value which coincides with the angular velocity of rotating machinery, the *stroboscopic effect* is likely to be produced and the shaft of the rotating machinery will appear to be stationary when it is actually moving. This can have disastrous consequences.

Fortunately there are some simple methods of preventing the stroboscopic effect from occurring, including:

- connecting adjacent single lamp luminaires to different phases of the electrical supply;
- using 'lead-lag' circuitry in the control gear of the luminaire which 'phase-shifts' the electrical supply between adjacent lamps in the same luminaire; and
- introducing a non-pulsating light source in the vicinity of the rotating equipment. Typically this is achieved with a tungsten filament lamp which is unlikely to give rise to flicker in the lamp output since the filament has considerable thermal inertia, which leads to good stability in the luminous output. Daylight is also a non-pulsating light source and the stroboscopic effect is much less likely to be a problem in workshops which have windows admitting natural light to the interior.

It is possible to put the stroboscopic effect to good use in some industrial situations, e.g. the adjusting and setting of the timing of the engine of a motor vehicle and the remote measurement of the speed of a rotating shaft. Figure 11.13 shows the production of the stroboscopic effect.

11.16 Light modulation

The light output from discharge lamps rises and falls in sympathy with the oscillations in the electrical supply due to its alternating nature, as shown in Figure 11.14. The effect can sometimes be perceived in fluorescent lamps.

An afterglow will be associated with the phosphors used in fluorescent lamps where the perceived ripple of constancy, or modulation of the light output, can vary depending upon the composition of the phosphors used, as shown in Figure 11.15.

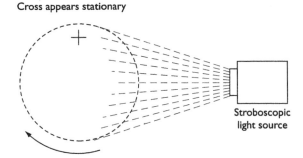

Cross appears stationary

Stroboscopic
light source

Disc rotates at speed which coincides
with pulse rate output of light from
stroboscopic source and therefore
cross appears to be stationary

Figure 11.13 Stroboscopic effect.

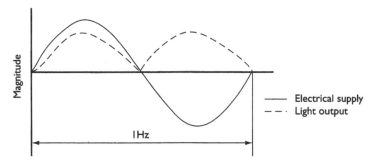

Figure 11.14 Light output from a.c. electrical supply.

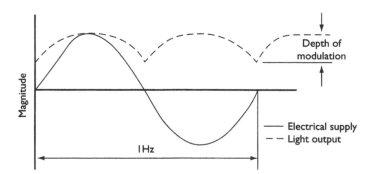

Figure 11.15 Light output modulation.

It will be evident that at the frequency values at which high frequency fluorescent luminaires operate (typically 30 kHz in the undimmed mode and between 30 kHz and 50 kHz throughout the dimming range), the effects described are unlikely to cause flicker problems.

11.17 Loss of perception

The loss of visual perception in individuals is often claimed to be a contributing factor in some workplace accidents. It is possible to subdivide loss of perception into groups, i.e. incorrect perception and total failure to perceive.

Incorrect perception can occur as a result of the following, either singularly or in combination: optical illusions, lack of modelling, light travelling from unusual directions and flickering light sources.

Failure to perceive can occur as a result of the following, either singularly or in combination: inadequate illuminance on the target, excessive glare in field of view and excessive luminance contrast.

11.18 Poor task lighting – effects on worker posture

When a particular visual target is viewed under either unsuitable or inadequate lighting, or both, workers will often readjust their posture in an attempt to improve their vision. Figure 11.16 shows the effect

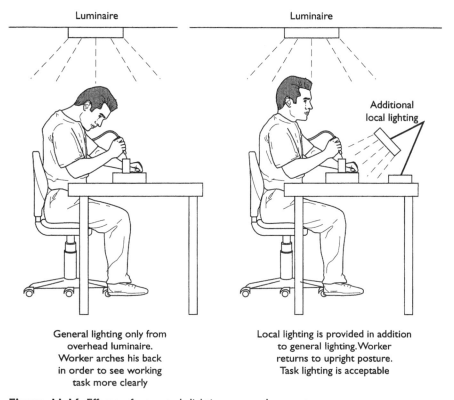

Luminaire	Luminaire

Additional local lighting

General lighting only from overhead luminaire. Worker arches his back in order to see working task more clearly

Local lighting is provided in addition to general lighting. Worker returns to upright posture. Task lighting is acceptable

Figure 11.16 Effects of poor task lighting on worker posture.

described. This may well produce associated health effects, e.g. musculoskeletal problems. In order to permanently improve the situation described it is important to give priority to task lighting. Failure to do so often produces excessive worker absenteeism which has an adverse 'knock on' effect for production output.

11.19 Effects of age of individual on task vision

It is well-documented that an individual will suffer changes in the eye which become more pronounced with age. These can be categorized as either 'perceptual' changes or 'physical' changes and are considered in Section 5.14.

If an interior is occupied by older individuals, it is beneficial to increase illuminance levels accordingly.

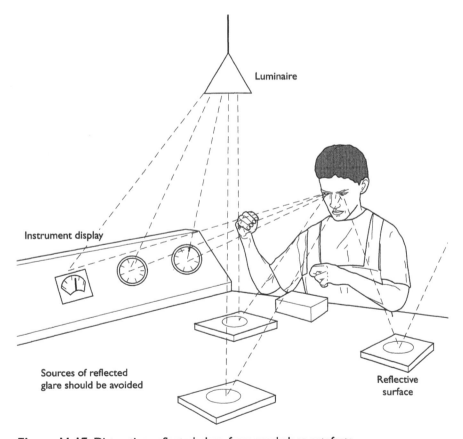

Figure 11.17 Distracting reflected glare from workplace artefacts.

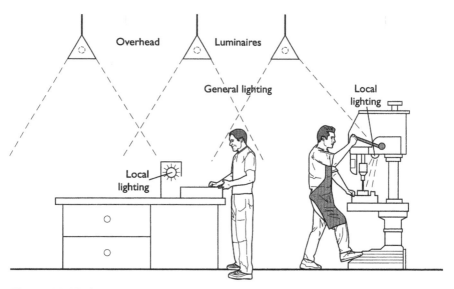

Figure 11.18 Optimum task lighting incorporating general and local lighting.

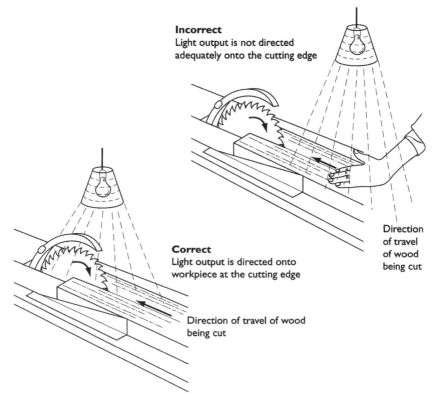

Figure 11.19 Importance of correct luminaire positioning.

11.20 Preferred task lighting strategies

Figures 11.17 to 11.23 inclusive show some examples of both recommended practice in the lighting of the workplace, and some of the effects produced when the task lighting is inappropriate.

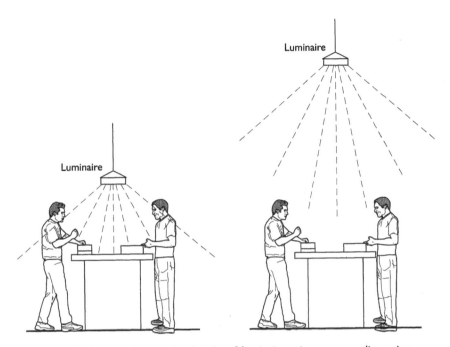

Figure 11.20 Increased mounting height of luminaires gives greater dispersion.

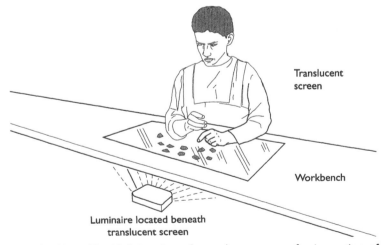

Figure 11.21 Use of backlighting through translucent screen for inspection of products.

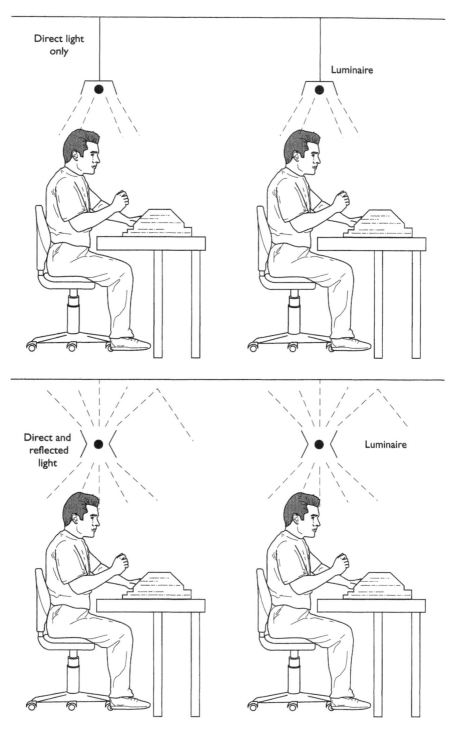

Figure 11.22 Combination of direct and reflected light leads to optimum visibility.

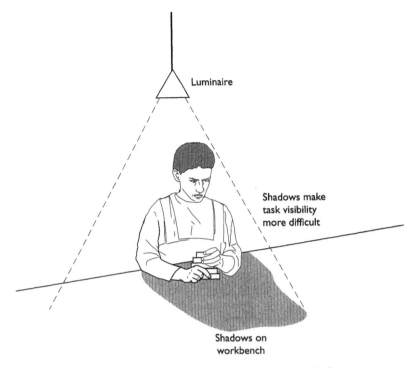

Figure 11.23 Incorrect luminaire siting causes shadows at the workplace.

11.21 Typical recommended illuminance values

Typical values of maintained illuminance for some workplace activities and locations are given in Table 11.2. A comprehensive schedule of recommended maintained illuminance values for many workplace applications and locations is given in the CIBSE Code for Interior Lighting 1994.[29]

Table 11.2 Typical recommended maintained illuminance values

Workplace	Typical recommended maintained illuminance values (lux)	
Engineering workshops	Rough e.g. bench work,	300 to 400
	detailed work,	700 to 800
	inspection	500 to 2000
Clothing and textile	Preparation,	300 to 400
	cutting,	750 to 850
	inspection	1500
Construction sites	Site roadways,	5 to 15
	general areas,	20 to 25
	crane loading	100 to 150
Electronics equipment manufacture	Component assembly,	300 to 1500
	printed circuit board work	500 to 750
	inspection	1000 to 1500
Food and drink	Abattoirs,	500 to 750
	bakeries,	300 to 500
	breweries,	300 to 750
	dairies,	300 to 500
	flour mills	300 to 500
Metal working	Iron and steel mills,	300 to 500
	foundries,	300 to 500
	inspection	500 to 1500
Furniture and timber	Sawmills,	300 to 750
	workshops,	300 to 750
	furniture manufacture,	300 to 750
	upholstery	500 to 1500
Glass production	General production,	300 to 750
	inspection	1000
Health care premises	General wards,	150 to 250 (day)
		3 to 5 (night)
	maternity wards,	200 to 1000
	laboratories,	300 to 750
	operating theatres	500 to 50 000
Paper and printing	Paper mills	300 to 750
	printing works,	300 to 750
	inspection	1000

Chapter 12

Lighting for specific industries and occupations

12.1 Introduction

It is not possible to include lighting applications which are applicable to all industries and occupations, but a selection is included which is considered to be representative.

12.2 Lighting for inspection

The inspection of products in a production process is a necessary element in monitoring the final quality of the finished article. The inspection may occur at various stages during the manufacture of a component, but will invariably include an inspection of the final article. The purpose of inspection, including visual inspection, is to identify any individual feature of the component which, if left undetected, would render it unsuitable for the purpose for which it is intended.

It will be evident that suitable and sufficient lighting contributes significantly to the overall process of visual inspection.

12.2.1 General inspection lighting techniques

The role of lighting for inspection is to improve the visibility of any features of a product which would subsequently lead to its rejection as a serviceable item. There is a multitude of manufactured products and each one of them will have a different combination of features which may prove defective in some way. It follows therefore that the optimum inspection lighting required will vary from product to product. It will be necessary to perform a visual task analysis in an attempt to achieve optimum task lighting conditions. Those factors which are to be considered in such an analysis include:

- The characteristics of the whole product which contains individual features which are required to be inspected, i.e. general shape of the product (two or three-dimensional), constitution of the product, e.g. transparent, translucent or opaque and surface finish qualities of the product, e.g. matt or specular.
- The individual features within the whole product which require inspection include form, detail and colour. Features of form are not generally restricted to small size features and often tend to be seen as part of the whole product, an example being inconsistencies in the bodywork finish of a motor vehicle which become more apparent when the whole vehicle body is viewed. Conversely features of detail are small and can normally be detected when only part of the product is being inspected, an example being a crack in the surface of some solid material. It will be evident that the size of a product will not usually influence colour features which require inspection.

Each individual inspection process has to be considered on its own merits and it follows that each process will require careful consideration in respect of the most appropriate lighting. However, as a general guide, features of form are usually best illuminated by luminaires with strong and clearly defined directional qualities of light output. Inspection of features of detail are normally revealed more easily when the illuminance levels are increased. The inspection of the surface colour features of a product is a process which is heavily dependent upon the spectral characteristics of the light sources used. Section 6.8 shows how the spectral power distribution of light reflected from the surface of an object is influenced by the combination of the spectral power distribution of the light source used and the spectral reflectance characteristics of the object being illuminated.

12.2.2 Detection of form defects

In general the form which a product appears to exhibit is strongly influenced by:

- the structural arrangement of the product;
- the manner in which it is lit; and
- the reflection characteristics of the material from which the product is constructed.

In combination these will lead to the development of three characteristic patterns:

- illuminance;
- highlight; and
- shadow.

The complexity of products and of manufacturing processes necessitate that lighting for discrimination of form be considered on its individual merits and as a consequence it would be unwise to suggest one method of inspection lighting which would be applicable in all cases. In many situations a 'trial and error' approach will often produce the best solutions.

12.2.3 Detection of detail defects

The ability of an individual to detect detail is heavily dependent upon:

* the characteristics of the human individual vision system;
* the size of the detail on a product; and
* the luminance contrast on the product.

An increase in the illuminance level will normally result in an increase in the sensitivity of the visual system. The size of the detail can be improved by magnification and luminance contrast will be improved by controlling the directional, reflective and transmission properties of the lighting.

12.2.4 Detection of colour defects

Where the discrimination of surface colour is critical, the apparent colour of the product is strongly influenced by the light sources used. The colour rendering of light sources is considered in detail in Section 6.8. The determination of the spectral composition of light reflected from an object, in terms of the spectral power distribution of the light sources used and the spectral reflectance properties of the surface being illuminated, are also considered in Section 6.8. BS 950[30] covers illuminants for colour matching appraisal.

12.2.5 Special techniques for inspection lighting

The techniques normally applied to inspection lighting include directional characteristics, reflection and transmission. There are, however, several special techniques which assist in the visual inspection process and these include:

* Stroboscopy: Inspection of equipment which under normal operation is rotating would necessitate stopping the rotation and this may be highly undesirable. The use of the stroboscopic effect (considered in detail in Section 11.15) in order to make the moving equipment appear stationary will avoid the need to stop the rotation, see Figure 12.1.
* Fluorescence: Many materials used in industry will fluoresce when subjected to ultraviolet radiation, i.e. the reaction produces visible radiation. An example of this is the coating of the surface of a product under inspection with fluorescent material. When the product is subjected to ultraviolet radiation, any surface defects will appear to be black.

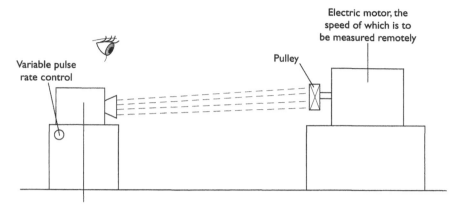

Stroboscopic light.
Output pulse rate is varied
until motor pulley appears stationary.
Speed of motor is then determined
from pulse rate setting of stroboscopic light

Figure 12.1 Remote speed measurement using stroboscopic effect.

- Polarization: With some transparent materials, polarized light can be used to advantage in the inspection process. The light output from a tungsten filament lamp is initially polarized and then transmitted through the product under inspection before a second polarizer is used. Any abnormalities in the stress pattern appearing from the second polarizer will indicate defects in the structure of the component.
- Fibre optics: Fibre optics can also be used to convey information about areas which would be visibly inaccessible, see Figure 12.2.

Scratches on polished metal surfaces can be revealed by using reflected light; the reflected light from the surface scratches will show up on a dark background. To inspect for surface flatness of products, it is possible to use monochromatic light from a low pressure sodium lamp in conjunction with optical flats; this creates 'optical fringes' which highlight any defects in the surface finish of the component.

12.3 Lighting for engineering workshops

It will be evident that suitable and sufficient lighting is essential. Since almost all workshops contain machinery which has moving parts, there is an obvious need to provide good task lighting which will illuminate potentially troublesome areas and the development and production of dangerously deceptive shadows must be avoided. It is equally important that glare should be avoided. Furthermore, rotating machinery is

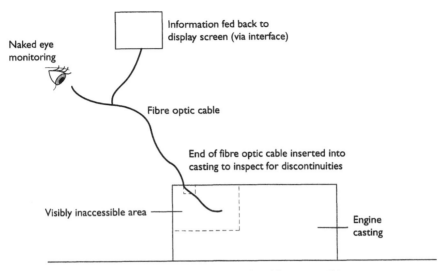

Figure 12.2 Use of fibre optics in inspection of visibly inaccessible areas.

capable of producing the stroboscopic effect if suitable preventative measures are not implemented. The concept of the stroboscopic effect is considered in detail in Section 11.15.

12.4 Lighting for textile and clothing manufacturing industries

Colour repeatability of style and shades is critical and it follows that light sources which offer good colour rendering index values must be used. *Metamerism* can be a problem if due consideration is not given. Metamerism is the term given to the phenomenon where two objects appear to have identical colours under one light source but under a different light source they appear to take on different surface colours. Colour matching lamps are often used where accurate colour comparison is necessary. Figures 12.3 and 12.4 show the concepts of both backlighting and frontlighting used in the inspection of textiles.

12.5 Lighting for building and construction sites

Work on construction sites is continually progressing and it will be evident therefore that task lighting should ideally be portable so that it is capable of being re-positioned as work progresses on site. If instantaneous light output is required then it will be evident that some discharge

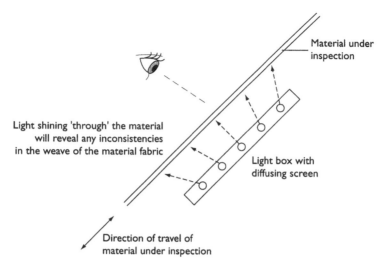

Figure 12.3 Use of backlighting in the inspection of materials.

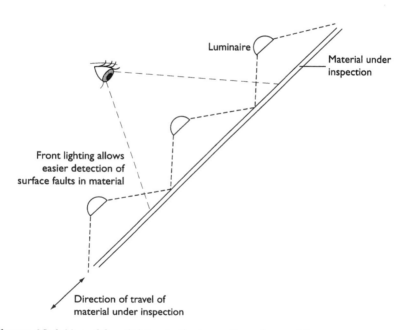

Figure 12.4 Use of frontlighting in the inspection of materials.

lamps will be inappropriate and in such cases incandescent lamps, typically tungsten halogen lamps, are often used. On some building and construction sites, the use of reciprocating and drilling machinery will cause localized areas where vibration will occur. Such vibration can lead to failure of incandescent lamps and this must be taken into consideration when planning a lighting installation.

12.6 Lighting for electrical and electronic equipment manufacture and assembly

The ability to detect colours accurately, e.g. the coding of electrical resistors, is essential and it will be evident that light sources used must have a good colour rendering index value. Local lighting is used extensively in the electronics industry although care must be taken to avoid producing veiling reflections on the displays of instrumentation. Electrical components which are made of brass or copper may not be easily recognizable under high pressure sodium lighting. Large parts of the electronics industry are miniaturized and in such situations high-powered magnifiers are often used. The magnifiers will improve the visual acuity of the workers and it will be necessary to provide correspondingly high illuminance values.

Where inspection of components and sub-assemblies is carried out, large low-luminance overhead luminaires may produce specular reflections which can be advantageous in the detection of some inconsistencies in the surface of the finished product. In some work areas where it is necessary to control the environmental conditions to precise values, the luminaires used must be of a type that will not adversely effect the environment.

Individuals with some colour defects are prevented from working in the electrical industry since their ability to correctly identify colour coded cables may be unacceptable. The restrictions on colour defectives being prevented from working in the electrical and electronics industry became less severe when the colour coding of cables was changed. The previously used system of red, black, green (for live, neutral, earth) was replaced with brown, blue, green/yellow (for live, neutral, earth). The system involving brown, blue and green/yellow coloured cable insulation allows more colour defective individuals to discern the difference between the conductors.

12.7 Lighting for food and drink industries

One major concern with the lighting of food and drink industries is the prevention of fragments of the lamp and/or luminaire from accidentally contaminating the food and/or drink during the processing. To this end, luminaires used in such environments must be so constructed as to prevent any possible contamination. Some processes involved in food and drink industries will produce corrosive atmospheres and luminaires used must be capable of normal operation in such environments. In order to satisfy food and hygiene regulations, many production areas have to be regularly hosed down with water and luminaires used in such environments must therefore be capable of withstanding hosing down.

12.8 Lighting for metal working industries

A major requirement in such industries is the avoidance of dangerously deceptive shadows which can be inadvertently produced in the vicinity of machinery. Disability glare can also become a problem in such industries where hot metal ingots also become high luminance sources. In order to produce an appropriate 'trade-off' in luminance values, the illuminance in the immediate surround to the ingots should be suitably controlled. With coloured moulding sands, used in some foundry processes, the use of high pressure sodium lamps may not be acceptable because of adverse colour discrimination effects.

12.9 Lighting for furniture and timber industries

In general, timber trades tend to produce quantities of combustible wood dusts, which can lead to the development of potentially flammable atmospheres. In addition treatment processes such as polishing and staining often use flammable solvents. In all cases, the luminaires used must therefore be capable of safe normal operation within the environments in which they are installed. BS 6467[23] covers the selection of electrical apparatus for use in the presence of combustible dusts.

12.10 Lighting for glass industries

Hot molten glass, as with hot molten metal, can produce high luminance sources which can lead to the production of potentially dangerous environments. In order to produce an appropriate 'trade-off' in luminance values, the illuminance in the immediate surround to the hot molten glass should be suitably controlled. Additionally many contaminants and vaporized materials are emitted during the production processes and luminaires used in such environments must be capable of operating safely within such atmospheres.

12.11 Lighting for premises used for health care

In areas where clinical examinations are carried out, some 'shades' of fluorescent lamp are prohibited. Those with specific blue components in their output can lead to an incorrect diagnosis of some patients where the blue tinge of the skin could inadvertently lead to the appearance of cyanosis. Fluorescent lamps which have a specified blue component in their output are often used in 'special care baby units' for the treatment of neonatal jaundice, see Section 7.14.

Lamps emitting heat should not be used in areas where exposed human tissue is likely, e.g. accident and emergency departments and operating theatres. All lamps and associated control gear must be silent in operation for hospital ward areas. Radio frequency interference and electromagnetic radiation interference should be eliminated so as to avoid causing interference to medical equipment, see Section 8.12.

12.12 Lighting for paper and printing industries

Some of the processes involved in the paper and printing industries can produce atmospheres which have an adverse effect on the performance of luminaires. Such agents include:

- damp
- dust;
- fibres;
- steam; and
- vibration.

Luminaires for use in the paper industries must therefore be capable of withstanding attack by such agents during their normal operation. Lighting in Printing Works[31] gives further information in respect of the lighting of such premises.

12.13 Lighting for mining

In early mining applications miners commonly suffered from a complaint referred to as *nystagmus*. The symptoms are an uncontrollable oscillation of the eyeballs, headaches, dizziness and a corresponding loss of vision. The root cause of such a condition is the working in very low illuminance levels for prolonged periods of time. It will be evident that early miners were particularly vulnerable to this complaint since they worked in very dimly lit environments and further that any available light was very poorly reflected from the surface of the coal. With the advent of electric lighting the condition has almost totally disappeared.

Light sources in underground applications in today's mines can be categorized as fixed or mobile. Tungsten filament lamps and low pressure mercury vapour (fluorescent) lamps are often preferred for fixed underground applications. Discharge sources other than fluorescent have been tried in underground applications but with mixed success. It will be evident that those lamp types which will not re-strike in the event of a momentary loss in electrical supply will present a problem in respect of the safety of underground workers. Luminaires used for mains-operated lighting have to satisfy stringent regulations on safety for use in the types of environments typically encountered in mines.

Mobile lamps are predominantly battery-operated cap lamps which are carried with the workers during their normal movements, some of which may be in areas where there is no mains lighting. A miniature tungsten lamp is often used in such equipment, which uses krypton as the fill gas, as described in Section 7.7.1.

12.14 Lighting for hazardous and hostile environments

An environment in which there exists a risk of fire or explosion is termed *hazardous*. An environment in which the lighting equipment may be subjected to attack from chemical, thermal or mechanical agents is referred to as *hostile*. All lighting equipment, including luminaires, must be capable of operating safely and efficiently in the environment for which it is designed and in which it is likely to be installed. To this end it is essential that lighting equipment itself must not be the cause of fire and/or explosion.

For explosions to occur, the presence of the following is essential:

- a flammable substance;
- an oxidizing agent; and
- a source of ignition.

Fuels are classified according to both their *group* and *T class*. The group to which a luminaire belongs is heavily influenced by the gases to which it is likely to be exposed. Details of groups are given in BS 5345.[22] The T class refers to a temperature classification whereby a maximum surface temperature is quoted which must be less than the ignition temperature of the gases and vapours likely to be present in the atmosphere. Dusts are fuels and it is important to be aware of the energy required to ignite a cloud of dust and the likelihood of a hot surface automatically igniting dust. BS 6467[23] covers electrical apparatus for use in the presence of combustible dusts. Sources of ignition include the closing or opening of electrical contacts, electrostatic discharges and friction.

Explosive atmospheres are categorized according to British Standard 5345[22] which is considered in Section 8.5.2.

Luminaires which use low pressure sodium sources are not permitted in areas where flammable gases and/or vapours are likely to be present, e.g. in petrochemical industries. Such concentrations could become accidentally ignited due to faulty electrical equipment or equipment which has inadequate protection against the environments in which it is to be used. Low pressure sodium lamps contain free globules of sodium which can cause combustion if it comes into contact with water.

12.15 Lighting for dockyards and shipbuilding

It is important to keep working areas free from obstructions and to this end the lighting of docks is usually achieved by means of luminaires installed some distance away from the dock edges. Dock-mounted floodlighting is often used and this can be complemented by floodlighting on board the decks of moored vessels.

Lighting equipment must be capable of withstanding the atmospheres which are likely to prevail in dockyards, which are typically saline.

It is also essential, in addition to providing suitable and sufficient task lighting, to make the edges of the docks as conspicuous as possible. This can be achieved by providing a high contrast between the dock edge and the water, often using a white painted strip along the edge.

Suitable and sufficient lighting must be provided for access, gangways and walkways, and tungsten halogen floodlights are often used for work on board ships. In addition emergency lighting must be provided in the event of the failure of the mains-powered lighting so that personnel involved in work activities below deck can escape safely.

Legislation is covered by the Docks Regulations[32] and COP 25[33] is the Approved Code of Practice with Regulations and Guidance.

12.16 Lighting for horticulture

Healthy plants manufacture sugars, a process which occurs only in the presence of light. Chlorophyll, which is a green pigment found in healthy leaves, makes possible this process, which is referred to as *photosynthesis*. Sugars provide the raw materials and the energy from which plant life ultimately manufactures cellulose. This breakdown of sugars to produce cellulose, a process termed *respiration*, occurs continually and if the rate of respiration is greater than the rate of photosynthesis, so that energy is used quicker than it is produced and stored, the plants will outgrow their strength.

Plants require adequate light (together with warmth, moisture, food and air) in order to stimulate growth and remain strong and healthy. Unfortunately, the lighting requirements of some plants will be markedly different from those of others in terms of such parameters as daily duration and intensity. Many plants require between ten and twelve hours of light each day for growth, and, for a considerable part of the year, lack of natural light is a limiting factor.

When artificial lighting is used to encourage the photosynthesis process, it is important to achieve a high level of intensity which should be commensurate with good natural daylighting. A tungsten filament lamp emits an excess of red wavelengths for full photosynthesis with the consequence that plants will develop long weak stems. Fluorescent lamps are more acceptable in terms of their spectral output.

12.17 Workers exposed to ultraviolet radiation

There are many items of equipment which emit ultraviolet radiation. Such equipment is used for a wide range of purposes including killing bacteria and producing fluorescence effects. In such situations the ultraviolet is deliberately produced. Workers may, however, be exposed to ultraviolet radiation as a consequence of unwanted by-products of industrial processes, e.g. arc welding.

Devices emitting ultraviolet radiation include:

- bacterial lamps;
- mercury lamps;
- carbon and xenon arcs;
- fluorescence equipment;
- plasma torches;
- welding equipment; and
- printing ink polymerization equipment.

The condition referred to colloquially as 'arc eye' or 'welder's flash' or more correctly photokeratitis occurs in an individual after looking with the naked eye at the arc produced in arc welding, see Figure 12.5. The symptoms are typically pain and discomfort similar to that experienced with grit in the eyes and an aversion to bright light. The cornea and the conjunctiva will show inflammatory changes. The severity of the condition is influenced by the duration, intensity and wavelength of the offending ultraviolet radiation. The sensation usually appears within a few hours of exposure and disappears after a few days although in all cases medical treatment should be sought.

Individuals involved in skiing and others likely to be involved in activities at high altitudes are particularly exposed to ultraviolet radiation since at elevated levels some of the ultraviolet from the sun has not been filtered by the atmosphere. Additionally the UV can be reflected from the surface of the snow.

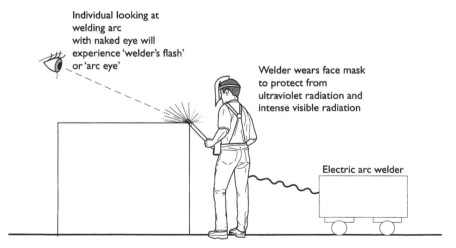

Figure 12.5 Effects of viewing electric welding arc with naked eye.

Chapter 13

Lighting and display screen equipment

13.1 Introduction

Display screen equipment is commonplace in industry, commerce and in domestic premises. Such equipment typically forms part of a workstation which may contain other types of office machinery. Most VDU workstations are designed based upon a seated individual but there are occasions where a standing position will offer some advantages for operators.

Working at display screens involves visual tasks and it is necessary therefore to perform some form of visual task analysis prior to designing a workstation. This will invariably involve some form of ergonomics assessment which takes into account, inter alia, the position of equipment, the possible position and likely movements of the operator, together with details of the time duration of such activities.

13.2 Radiation associated with the use of display screen equipment

Radiation is described as the emission or transfer of radiant energy. It will be evident that the emission of radiant energy in the form of light is the primary purpose for which the display screens are used. It is often the situation, however, that there is an attendant emission of associated by-products from display screen equipment including:

- ultraviolet radiation;
- infrared radiation;
- heat;
- radio waves; and
- x-rays.

Visible light can influence individuals in a positive manner but some forms of emissions of energy can have a negative or deleterious biological effect on humans, particularly when the intensity values are high and/or the duration of exposure is long.

Optical radiation includes ultraviolet, visible and infrared radiation, as described in Section 2.2. Visible radiation is emitted at relatively moderate intensity values, compared to that emitted by typical room surfaces; however, ultraviolet radiation is intercepted by the glass of the face of the cathode ray tube (CRT), or in other equipment it is unlikely to be emitted. Levels of both ultraviolet and infrared radiation emitted by CRTs are likely to be considerably below occupational exposure limits.

It is well documented that CRTs are sources of X-rays. Tubes and associated equipment are designed so as to restrict the emission levels to well below the occupational exposure limits. It has to be appreciated that radiation emitted by a source will only be detected if its magnitude is greater than that emitted by the sources in the background. With X-rays, the background level is normally provided by a combination of cosmic radiation and by radiation from other radioactive materials within the surrounding buildings and the ground. It is generally accepted that X-rays do not emit levels exceeding that of background materials.

13.3 Screen image

Of necessity, the information displayed on a screen is seen as variations in luminance. As a consequence the luminance contrast will influence the visibility of the text on the screen. Furthermore the luminance of the visual field surrounding the screen is unlikely to be constant and this will have an effect on the overall conditions prevailing.

Luminance imbalances may occur in areas where display screen equipment is used. These imbalances can be conveniently sub-divided into either static and dynamic.

- *Static luminance imbalance* (SLI) is likely to occur when there are widely-varying luminance values in the line of sight of the display screen equipment operator. One of the more frequent causes of static luminance imbalance is when a display screen is positioned against a background of an excessively high luminance source, e.g. a window. Figure 13.1 shows the effect of the static luminance imbalance (SLI) described.
- *Dynamic luminance imbalance* (DLI) is likely to occur when a display screen equipment operator views objects with markedly-differing luminance values in quick successive movements. If, for example, an operator changes gaze from a scene of low luminance to a different scene which has a relatively high luminance, then the visual system will be constantly making changes in adaptation. Figure 13.2 shows the effect of the dynamic luminance imbalance (DLI) described.

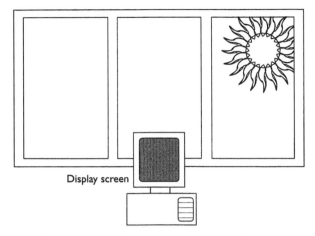

Figure 13.1 Static luminance imbalance. Luminance of display screen is much lower than that likely to be experienced through the window, a scene which may include bright sunlight.

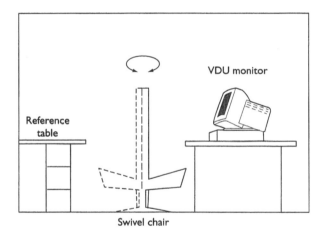

Figure 13.2 Dynamic luminance imbalance. VDU operator swivels from viewing the VDU screen to reading information on the reference table. The two scenes may have widely differing luminance values, leading to the development of dynamic luminance imbalance.

In Figure 13.3 the luminaire position is such that it will produce an out-of-focus image of it in the screen. This throws a veil of light in front of the screen, which accounts for the term veiling reflections. The out-of-focus image will result in an additional problem where the operator is likely to experience irritating focal length variations.

The viewing distance between the operator and the display screen is typically 50 cm. Figure 13.3 shows that the distance between the operator

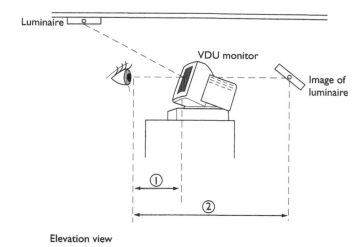

Elevation view

Figure 13.3 Focal length variations associated with veiling reflections. 1 Distance between operator and screen. 2 Distance between operator and image of overhead luminaire.

and the apparent position of the luminaire as seen by the operator in the screen is far greater. The operator will attempt to view both the screen and the out-of-focus image of the luminaire which will produce visual problems including fatigue.

Several terms are used to describe the characteristics of the image seen on the screen of a visual display unit.

13.3.1 Resolution

This is the term used to describe the smallest measurable detail contained in a visual presentation. The maximum number of lines that can be displayed on a display screen is referred to as the resolution.

13.3.2 Flicker and refresh rate

The images displayed on screens are not persistent images as would be the situation with information printed on paper. They take on the appearance of a persistent image by virtue of the characteristics of the human vision system. If, however, the image on the screen is not refreshed constantly then the image will not appear persistent and a flickering sensation will be detected. Flicker can be described as the perception of a time-varying brightness.

It will be evident that if a flickering sensation is perceived by individuals, they are likely to suffer from some forms of visual maladies, e.g. headaches, eyestrain. Refresh rates which are capable of being varied between 70 Hz and 90 Hz are typically provided on displays.

13.3.3 Jitter

This is a term applied to the spatial instability of an image. In practice, this means that the image is not aligned on the same point of the screen at successive refreshes of the picture. It is difficult to separate the perception of flicker and the perception of jitter. It is possible that jitter is caused by interference from adjacent equipment or similar devices which rely, for their operation, upon the production of magnetic fields.

It will be evident that the presence of jitter on a screen will create visual problems for the operator of display screen equipment.

13.3.4 Screen contrast

Luminance contrast indicates the relationship between a visual target and its corresponding background. Good contrast is essential when considering both readability and legibility of text contained in displays. It will be evident that the poor contrast of a screen display will create visual problems for the operator.

13.3.5 Active and passive displays

An example of active displays are those displays on the screen of visual display units whereas an example of passive displays is those produced by liquid crystal displays (LCDs).

13.3.6 Cathode ray tubes (CRTs)

The glass surface of the screen will form a reflecting medium and objects within the interior may be reflected, thereby producing unwanted 'out-of-focus' images in the screen. The magnitude of illuminance in the interior may adversely effect the contrast on the screen.

13.3.7 Liquid crystal displays (LCDs)

Reflections on LCD displays are typically less than those on CRT displays since LCD displays often have flat surfaces rather than curved surfaces. LCDs lose contrast under poor illuminance levels.

13.4 Glare control on VDU screens

The control of screen glare is often best achieved by using a 'trial and error' approach, having considered three basic factors:

• choosing a suitable location for the screen with respect to all potential glare producing sources;

- choosing suitable equipment, including any possible devices;
- choosing suitable and sufficient lighting.

The use of *positive polarity screens*, i.e. dark characters on a light background, is often considered beneficial, as is the treatment of the glass screen with an anti-glare coating.

Ergonomists may suggest the use of hoods on screens in environments where there are likely to be several high luminance sources present, although it could be argued that if hoods are required then there is the strong possibility that there are problems generally with the lighting installation.

13.5 Lighting for VDU areas

Careful forethought in the planning of a lighting installation in interiors where display screen equipment is in use can contribute greatly to the acceptability of the overall visual environment.

To this end it is essential to appreciate that when designing a lighting installation for areas containing display screen equipment, with the singular intention of providing sufficient and suitable lighting for the screens only, there is a strong possibility that other visual problems elsewhere within the interior are likely. It will be evident that other visual tasks need to be carried out in the same office space and at the same time.

When a display screen equipment operator is viewing a screen, other parts of the interior will be within their visual field and this will lead to problems with adaptation. Light scattered in the eye will reduce the contrast of the image subsequently formed on the retina. The effect of this is to produce impaired vision. In addition if the display screen operator momentarily glances away from the screen, transient changes in adaptation will occur and again vision can be impaired.

There are several options available to the designer in an attempt to produce optimum lighting conditions. The most simple involves a re-positioning of one or more of the following:

- the light source;
- the screen; or
- the operator.

It will be evident that any re-positioning of the light source is the least practical since in many cases luminaires are ceiling-mounted in fixed locations. There are however much simpler remedies, for example the ergonomics of the operator is highly significant. In Figure 13.3 the offending luminaire is shown as being located in parallel with the front of the screen. It follows that the operator will see the largest possible image, i.e. the full length of the offending luminaire. Rotating the screen through 90° will ensure that the projected area of the offending luminaire is at a minimum where the operator will only see the 'end' of the luminaire. It is possible however that this manoeuvre may result in the operator's line of sight being at 90° to the normal to a window wall, which conflicts with recommendations (see Section 13.6). In such situations it may be worth considering relocating the offending luminaires through 90°.

In interiors where the installation of uplighters is inappropriate then downlighters can be used. In such situations restrictions are placed upon the luminous output characteristics of the luminaires. In an attempt to avoid high luminance reflections appearing on display screens it is important to use luminaires with an appropriate luminous intensity distribution, which subsequently limits the luminance seen by screen operators.

If the relationship between the terminal and operator could be established then it would be possible to determine the appropriate geometry, which in turn would allow calculation of luminaire characteristics necessary to prevent direct vision of the light sources. However such information is seldom available due to the wide range of tasks and screen types in use.

The work activities of those who use display screen equipment are covered by both EC and UK legislation, which specifies minimum standards for the visual environment, including the lighting conditions. In order to satisfy the legislation, display screen equipment tasks must be classified in accordance with the conditions specified therein. The tasks can influence the magnitude and severity of reflections likely to be encountered on the screen. The three categories relative to areas where display screen equipment is used are designated Category 1, Category 2 and Category 3. Downlighter luminaires for use in each of the three areas specified are known as Category 1, Category 2 and Category 3 luminaires. For each of these luminaire categories, the luminance above a predetermined angle, referred to as the *critical angle*, is limited to 200 candela per metre^2. The values of critical angles referred to, for Categories 1, 2 and 3 luminaires, are 55°, 65° and 75° respectively. Figure 13.4 shows the geometry of the critical angles described.

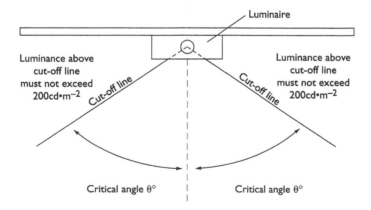

Figure 13.4 Critical angles for downlighter luminaires used in VDU areas.

Category 1 luminaires are used typically where there is a high number of screens, in an area where extensive use is likely and where errors occurring as a consequence of misreading the data on the screen are unacceptable. Category 1 luminaires are typically used in air traffic control rooms and emergency services control rooms. Category 2 luminaires tend to be used far more than other category luminaires and their use is typically in areas where display screen equipment is located at each workstation. Category 3 luminaires are usually used where screen usage is random and where the number of screens is relatively low.

Due to the limits placed upon the intensity distributions, Category 1 luminaires are less efficient than Category 2 luminaires, which in turn are less efficient that Category 3 luminaires. In addition to the requirement to amend the spacing of luminaires in accordance with their light intensity distributions, it is possible with such installations to produce an impression of a dimly-lit and depressing environment. This is due to the distribution of light at such tight angles which produces a high luminance contrast between the working plane and the room fabrics. In an attempt to avoid the creation of a gloomy room appearance, the use of an indirect lighting component will be an advantage.

The Health and Safety (Display Screen Equipment) Regulations 1992,[34] under the Health and Safety at Work Act 1974,[35] refers to minimum health requirements in relation to work with display screen equipment. Employers must take steps to ensure that all workstations under their control comply with the Regulations. In general terms the Regulations require lighting conditions which are satisfactory and which ensure that an appropriate contrast between display screen equipment and the background environment is provided. Any glare or distracting screen reflections must be prevented. Guidance on the Regulations (L26)[36] gives additional information.

13.6 Optimum conditions for visual comfort in display screen equipment operators

It is convenient to categorize methods of avoiding problems with display screen equipment into three, i.e. screen, keyboard and the environment.

In respect of the screen, the contrast should be adjustable and the screen should be capable of being tilted. Ideally display screen equipment should be inclined at an angle of typically between 30° and 40° below the operator's horizontal, as shown in Figure 13.5. It should also be possible to fit glare reduction devices to the screen.

The keyboard should be kept separate from the screen to allow freedom of comfort for the operator and the keys should be cleansed frequently to prevent them from becoming shiny thereby developing high luminance sources.

When seated at a terminal, an operator's line of sight should ideally be parallel to the window wall, and furthermore a terminal should be located as far from the window as possible. Clearly these recommendations only

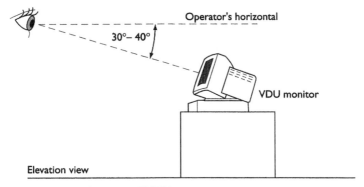

Figure 13.5 Suggested inclination of VDU monitor.

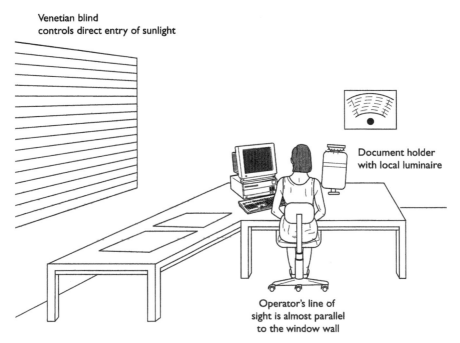

Figure 13.6 Typical layout of operator and VDU screen.

apply to interiors with windows on one wall only. The illuminance in proximity to a display screen can be as low as 100 lux to 150 lux, although in order to allow reference documents to be read without difficulty, the illuminance range 300 lux to 500 lux is normally recommended. Figure 13.6 shows a typical recommended layout.

It will be evident that in an attempt to accurately assess and prescribe suitable remedial action when striving for optimum visual comfort conditions, it is essential to use an approach which combines many factors.

Visual fatigue and general eye discomfort can occur as a consequence of malfunctions in the visual system and/or inappropriate corrective lenses. It can also be attributed to the workstation being used and furthermore it can be caused as a result of working practice anomalies, e.g. boredom, monotony. The lack of suitable and sufficient lighting, with specific attention to directional qualities, is another possible cause of eye discomfort.

Chapter 14

Lighting surveys

14.1 Introduction

Lighting surveys are often undertaken in response to adverse criticism from workers who complain that the lighting in their workplace is in some way unsuitable and/or insufficient. Whilst such an allegation may sometimes be justified, it is often the case that in an environment where there are reported visual problems the cause is non-lighting in origin.

Such non-lighting influences on the visual environment must therefore be investigated before any decision is taken to implement a full lighting survey. Typical non-lighting causes include:

- changes in the condition of the decor, e.g. room fabrics have become dirty and their corresponding reflectance values have decreased;
- changes in condition of fenestration, e.g. windows have become dirty and therefore the transmission of natural daylight to the room interior will be impeded;
- changes in layout of the interior, e.g. workstations have been re-positioned; and
- changes in working practices and activities.

If it is found that non-lighting factors contribute to the adverse visual environment then they must be remedied, after which further opinions should be sought from the workforce as to their effectiveness.

If it is evident that non-lighting factors are not wholly and exclusively responsible for the visual conditions experienced, then it is likely that a full lighting survey will reveal other possible causes which will need to be addressed.

14.2 Instrumentation

The major parameters of interest likely to be considered in a lighting survey include illuminance, luminance, reflectance and daylight factor.

Equipment is available which is capable of displaying, either in analogue or digital form, more than one of the parameters listed.

14.2.1 Illuminance measuring equipment

Such instruments typically incorporate a selenium or silicon photovoltaic cell. The specification of the instruments typically includes details of:

- spectral response;
- angular response;
- linearity of response; and
- operational characteristics of instrument in adverse temperature conditions.

The spectral response of the cells typically used in the construction of illuminance measuring instruments will differ from the response of the human visual system. It is therefore necessary to correct for this differential by applying some form of compensation. This is often achieved by using filters and when such filters are incorporated into an instrument, the device is referred to as *colour corrected*.

The magnitude of illuminance recorded by, and displayed on, an instrument, is influenced by the cosine of the angle the incident ray subtends with the normal to the plane of the instrument detector. It is necessary therefore to apply some form of compensation which takes into account the direction of the incident light falling upon the light-sensing detector. Instruments which are capable of measuring illuminance values accurately from any incident direction are referred to as *cosine corrected*.

The electronic components incorporated into the device will influence the linearity of response of the measuring equipment and the equipment may also be affected by adverse temperature conditions.

14.2.2 Luminance measuring equipment

Luminance measuring instruments incorporate photovoltaic cells which also require to be colour corrected and which are required to have a linear response.

14.2.3 Daylight factor meters

Instrumentation is available which can measure directly values of daylight factor.

14.3 Survey methods

Various methods are available for conducting a full lighting survey. Most of them rely upon the same basic principles.

14.3.1 Determination of the minimum number of measuring points

For any lighting survey to yield a meaningful representation of the actual lighting conditions applying in an area the survey must be based upon a minimum number of measuring points. It has to appreciated that the minimum number of measuring points decided upon may be justifiably increased. Measurements at fewer points than this minimum number may yield a totally unrepresentative picture of the lighting conditions applying in an area.

One method used for calculating the minimum number of measuring points initially involves the calculation of a parameter known as the room index, as detailed in Section 9.10.

Following calculation of the room index value, a second parameter, 'x', is calculated. For all values of room index, the parameter 'x' is taken as the next highest integer, with the limit that for all values of room index equal to or greater than 3, 'x' is taken as the value 4.

The minimum number of measuring points is then calculated from the expression:

$$\text{Minimum number of measuring points} \;=\; (x + 2)^2 \qquad (14.1)$$

Table 14.1 gives corresponding values of room index, parameter 'x' and the minimum number of measuring points. All measuring points are normally taken at 0.85 metres above ground level for routine horizontal tasks, where 0.85 metres is taken as the height of the top of a desk.

Table 14.1 Examples of values of room index, parameter 'x' and minimum number of measuring points

Room index value	Parameter 'x'	Minimum number of measuring points
0.8	1.0	9
1.8	2.0	16
2.0	3.0	25
2.8	3.0	25
3.0	4.0	36
3.6	4.0	36
4.2	4.0	36

14.3.2 Presentation of information

The presentation of the information obtained from a lighting survey can take many forms. One method involves a combination of tabulated and pictorial representation. The initial information can be listed as follows:

- Location of survey, e.g. general administration room No. 2.
- Date of survey.
- Name of person carrying out the survey.
- Reasons for the survey, e.g. complaints from the workforce.
- Visual tasks, e.g. VDU operation, proof reading.
- Visual planes, e.g. horizontal.
- Recommended illuminance values for area to be surveyed, typically obtained from CIBSE Code for Interior Lighting.[29]
- Room dimensions, e.g. length, height and width.
- Window dimensions, e.g. height and width.
- Total glazed area.
- Ratio of glazed area to floor area.
- General condition of room surfaces, e.g. clean, moderate, dirty.
- Luminaire types – lamp types and date of last lamp change.
- Ambient room temperature.

It is advisable to have scale drawings of the interior to be surveyed which will enable a layout of the existing lighting installation to be recorded. In many situations, photographs showing sites of particular concern can be extremely useful where scale drawings might not allow a true represen-tation of the conditions applying in particular areas. Information

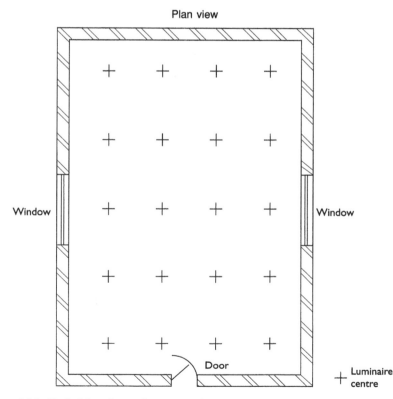

Figure 14.1 Typical interior to be surveyed.

presented on scale drawings can appear 'cluttered' and it may therefore be beneficial to use symbols and abbreviations on the scale drawings which allow full information to be presented separately in tabular form. Figure 14.1 shows a typical interior to be surveyed. The mounting height, i.e. between the horizontal centre line of the luminaires and the working plane, is 2.5 metres and the major room dimensions are 15 metres by 10 metres.

From the calculations detailed in Sections 9.10 and 14.3.1:

$$\text{Room index} = \frac{15 \times 10}{2.5\,[15 + 10]} \qquad (14.2)$$

$$= 2.4$$

It follows that the value of parameter 'x' is 3, giving a minimum number of measuring points of 25.

Figure 14.2 shows the position of 25 measuring points, each of which is given an alphanumeric reference. Table 14.2 shows a suggested schedule of illuminance readings which takes into account contributions from both daylight and artificial lighting.

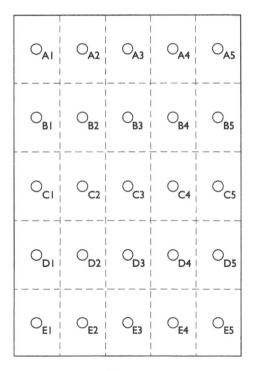

Plan view

Figure 14.2 Layout of survey measuring points. ○ measuring point location (measurement made at working plane level).

Table 14.2 Schedule of illuminance readings

Measuring point location	Measured internal illuminance (lux) due to both daylight and artificial light
A1	455
▼	▼
▼	▼
B2	495
▼	▼
▼	▼
C3	515
▼	▼
▼	▼
D4	460
▼	▼
▼	▼
E5	505

If daylight is thought to be a contributory factor in adverse visual environmental conditions experienced then it is possible to include its effects into the survey. Consider Figure 14.1 which shows a typical interior in which daylight and/or artificial lighting may be contributory factors in the visual environmental conditions. A step-by-step procedure allowing contributions from both daylight and artificial light to be separated is detailed in Section 14.3.3.

14.3.3 Method of separating illuminance contributions due to daylight and artificial lighting

1. Decide upon the locations of the measuring points.
2. Use a recording illuminance meter located external to the interior under survey, and ensure that its location is such that the readings of external illuminance obtained will not be influenced by the artificial lighting operating within the interior. Take external illuminance readings which must be time referenced.
3. Take and record illuminance readings at the agreed internal measurement points with the artificial lighting switched off, recording the time at which individual measurements are taken. The illuminance readings obtained will be due to daylight only.
4. Calculate the values of daylight factor at each of the agreed measurement points and record these values as shown in Table 14.3. The calculations are based upon the readings described in steps 2 and 3 which must be time-synchronized. Daylight factor measurement is discussed in Section 7.3.
5. Artificial lighting should then be switched on and allowed to warm up for a period of at least 15 minutes.

Table 14.3 Calculation of daylight factor values for lighting survey

Col. ① Measuring point location	Col. ② Time	Col. ③ Measured external illuminance (lux) due to daylight	Col. ④ Measured internal illuminance (lux) due to daylight	Col. ⑤ Daylight factor (%) $\dfrac{(Col. ④) \times 100}{(Col. ③)}$
A1	1100	8500	187	2.2
▼ ▼	▼ ▼	▼ ▼	▼ ▼	▼ ▼
B2	1115	8600	198	2.3
▼ ▼	▼ ▼	▼ ▼	▼ ▼	▼ ▼
C3	1130	8400	210	2.5
▼ ▼	▼ ▼	▼ ▼	▼ ▼	▼ ▼
D4	1145	8700	122	1.4
▼ ▼	▼ ▼	▼ ▼	▼ ▼	▼ ▼
E5	1200	8400	160	1.9

6. Take and record illuminance readings at the same interior measuring points. The illuminance readings obtained will therefore contain contributions from both natural daylight and artificial lighting. It will thus be possible to separate the contributions due to daylight only, from which the resulting contributions due to artificial light sources only can be deduced as shown in Table 14.4.

The step-by-step method discussed allows separation of the lighting contributions so that it will be possible to consider the effects of each in isolation and thus take any remedial action considered necessary.

In all surveys involving an interior lit by fluorescent sources it is essential to monitor the prevailing ambient temperature. Low ambient temperature will adversely affect the luminous output from fluorescent sources, as discussed in Section 7.8.2.

Illuminance and luminance values should also be taken on the room surfaces, i.e. floor, walls and ceiling and a mean value calculated for each of the surfaces. In order to strive for optimum visual conditions, preferred illuminance ratio and luminance ratio values should be in accordance with the values shown in Figures 9.3 and 9.4.

14.4 Interpretation of the data obtained

From the data obtained and recorded in Tables 14.2, 14.3 and 14.4, it will be possible to calculate values of minimum illuminance and mean

Table 14.4 Separation of contributions due to daylight and artificial lighting

Col. ①	Col. ②	Col. ③	Col. ④	Col. ⑤	Col. ⑥
Measuring point location	Time	Measured external illuminance (lux) due to daylight only	Inferred internal illuminance (lux) due to daylight only (Col. 3 x DF)	Measured total internal illuminance (lux) due to both daylight and artificial lighting	Inferred internal illuminance (lux) due to artificial lighting only (Col. 5–Col. 4)
A1	1300	8000	176	552	376
▼	▼	▼	▼	▼	
▼	▼	▼	▼	▼	
B2	1315	8200	139	596	457
▼	▼	▼	▼	▼	
▼	▼	▼	▼	▼	
C3	1330	8150	204	615	411
▼	▼	▼	▼	▼	
▼	▼	▼	▼	▼	
D4	1345	7950	111	545	434
▼	▼	▼	▼	▼	
▼	▼	▼	▼	▼	
E5	1400	8150	155	580	425

N.B. DF = daylight factor obtained from Table 14.3.

illuminance. From these values it is possible to calculate the uniformity ratio, as shown in Section 11.12. Values of uniformity ratio less than 0.8 will often lead to visual problems being experienced within an interior. If the values of illuminance ratio and luminance ratio differ markedly from those shown in Figures 9.3 and 9.4, problems are likely to develop.

14.5 Recommendations

Lighting surveys are performed usually following complaints from the workforce in connection with an adverse visual environment, and then only when non-lighting causal factors have been investigated and eliminated.

Recommended remedial action, based upon the findings of a lighting survey, can vary from simply replacing lamps which may be nearing the end of their life expectancy, to a full reappraisal of the effectiveness of the existing installation.

Luminaire cleaning and room surface cleaning are items which should ideally be included in a cyclic maintenance programme.

If, however, a major reappraisal of the existing lighting installation is suggested, then caution should be exercised. If this conclusion is reached as a consequence of illuminance and luminance measurements taken, then it has to be appreciated that implementing such far-reaching action should not be based upon the outcome from one set of measurements only. It is possible that one set of readings could contain spurious values, all of which could lead to conclusions which are not necessarily representative of the true situation. In order to produce conclusions in which a degree of confidence can be held, repeat sets of readings are essential.

Chapter 15

Legislation

15.1 Overview

Different countries throughout the world have their own legislation which must be adhered to so as to produce safe working environments. Some countries produce national legislation whereas in others the laws relating to health and safety issues (including lighting) vary from state to state or between provinces.

It would, therefore, be beyond the scope of this book to list, and consider in detail, all global lesgislation relating to lighting in the workplace. Readers are advised to seek the assistance of the appropriate authorities in order to ensure compliance with legislation.

Legislation applicable to lighting within the United Kingdom is considered in Appendix A.

Legislation in the United Kingdom

A.1 Introduction

In the United Kingdom the foundation for health and safety legislation is the Health and Safety at Work etc. Act 1974.[35] This Act sets out general duties which employers have towards both employees and members of the general public. Furthermore the Act sets out the general duties which employees have to themselves and to each other.

The principle of 'so far as is reasonably practicable' allows the duties to be qualified in the Act. It follows that the degree of risk associated with any workplace activity needs to be balanced against the time taken, trouble caused, likely cost and physical difficulty in implementing such measures in order to avoid or reduce the risk.

More explicit information in respect of the requirements placed upon employers so that they manage health and safety is given in the Management of Health and Safety at Work Regulations 1992[37] together with the Management of Health and Safety at Work: Approved Code of Practice (L21)[38] (see Section 15.3.1).

Much of the United Kingdom's health and safety legislation has its origins in Europe. Generally proposals from the European Commission (EC) may be accepted by member states who, in turn, will then be responsible for making the proposals part of their own national legislation.

The author and the publishers make no warranty, express or implied, nor assume any liability in respect of the use, or subsequent damages resulting from the use, of the information contained in this chapter. Furthermore, compliance with the recommendations given in the following sections does not guarantee compliance with the specified legislation but implementing the recommendations in the guidance should help reduce the probability of contravention arising.

A.2 Acts, regulations and approved codes of practice

Regulations are law which has been approved by Parliament. Regulations are usually made under Acts of Parliament as a consequence of proposals originating from the Health and Safety Commission. The Health and Safety Executive is the operating arm of the Health and Safety Commission.

Approved codes of practice (ACOPs) provide examples of good practice and are designed to give advice, although employers are at liberty to take alternative measures providing that they achieve the requirement of 'reasonably practicable'.

Approved codes of practice are referred to as having a special legal status. If an employer is prosecuted for a contravention of health and safety legislation and it is subsequently proved that the employer had not complied with the relevant provisions of the appropriate code of practice, then a court may find the employer at fault, unless the employer can show that the law has been complied with in some other way.

In some workplace activities some risks are considered so great or control measures considered so costly, that it is deemed inappropriate to allow employers discretionary powers in deciding how best to eliminate the risks. In such situations, regulations identify the risks and as a consequence identify specific remedial action which must be taken.

Some regulations apply to all workplace activities, whereas other regulations apply to workplace hazards which are relevant to specific industrial activities.

A.3 The law relating to lighting

Many regulations and approved codes of practice contain references to lighting.

A.3.1 Management of Health and Safety at Work Regulations 1992[37]

Regulation 3 refers to risk assessment. It requires all employers, including those who are self-employed, to assess the risks to workers and any other individuals who may subsequently be affected by their actions. Employers with five or more employees must make a record of the outcomes of their risk assessment.

In many cases employers may already carry out risk assessment as part of their daily duties, and, as a consequence, they will become aware of any malfunctions as they occur and take necessary remedial action. The framework of this regulation requires that such assessment should be formalized and that employers should keep records of the findings of such assessments.

L21 Management of Health and Safety at Work: Approved Code of Practice[38] gives additional information. The management of health and safety at work involves an element of lighting.

A.3.2 Workplace (Health, Safety and Welfare) Regulations 1992[39]

Regulation 8 refers to lighting. Lighting should be sufficient to allow individuals to go about their normal working practices, use workplace facilities and be able to move around the work area in safety and without experiencing any eyestrain.

Stairways should be lit in such a manner so that potentially dangerous shadows are not cast over the stair treads. When required, local lighting should be installed at workstations and at points where there is the potential for problems to occur. These locations are not restricted to internal applications and can equally apply to external applications.

Luminaires and light sources which cause dazzle or glare should be avoided and further they should be positioned so that they do not constitute a hazard. Switches controlling the lighting should be positioned so that they do not constitute a risk to persons who have to operate them.

The luminous output from luminaires should not be obstructed by equipment which is stored and/or stacked in such a manner that the light reaching the working areas is seriously reduced. All lighting equipment should be maintained regularly, e.g. luminaires cleaned and lamps changed on a cyclic basis in accordance with manufacturers' recommendations. If any part of the lighting equipment becomes dangerous it should be withdrawn and replaced immediately.

In respect of the use of natural lighting, windows and skylights should be cleaned at regular intervals. Furthermore they should be kept free from any obstructions, e.g. stored and/or stacked materials which would restrict the maximum possible transmission of daylight into an interior. A suitable balance should be obtained in respect of the heat transfer and glare from the admission of daylight into an interior. Workstations should be located within interiors so that they take full advantage of the daylight being admitted. Occasionally natural lighting is not possible where, for example, windows are covered so as to deliberately prevent the interiors from being viewed from the exterior for security reasons.

Emergency lighting should be provided in workplaces where the sudden loss of artificial lighting leads to a serious risk to individuals. Examples of such situations include chemical process plants where shutdown procedures must be followed, necessitating manual control which requires lighting in order for the process to be carried out. Emergency lighting should be supplied from a power source independent of that supplying normal lighting. In the event of a failure of the supply to the normal lighting, the independent power source supplying the emergency lighting should be activated automatically. The lighting provided by the emergency system should be sufficient to allow individuals at work to take any action considered necessary so as to ensure the health and safety of themselves and others.

A.3.3 Health and Safety (Display Screen Equipment) Regulations 1992[34]

The general duties require employers to take into account the working conditions and risks involved in the workplace when selecting equipment, ensuring that it is suitable for the use to which it will be put, ensuring that it is maintained to a satisfactory standard and providing adequate information, instruction and training for those who will use the equipment.

Any room lighting or task lighting which is provided shall be such that it ensures satisfactory lighting conditions together with an appropriate contrast between the screen and the background environment, bearing in mind the type of work undertaken and the vision requirements of the operator or user.

Any possible disturbing glare and screen reflections shall be prevented by the appropriate coordination of the workplace and the workstation layout. Due consideration must be taken of the location and properties of the artificial light sources used.

In respect of reflections and glare, workstations must be designed so that any sources of light will not cause glare or distracting reflections on screens. Such sources of light include windows, openings, walls and fixtures. To this end windows shall be such that it is possible to control the quantity of daylight falling onto a workstation, e.g. by the use of blinds.

Regulation 5 refers to an employer's liability in respect of the provision of eye and eyesight tests for employees who use display screen equipment.

The Guidance on the Regulations (L26)[36] gives additional information.

A.3.4 Health and Safety (Safety Signs and Signals) Regulations 1996[40]

The light output emitted by any equipment must produce a luminous contrast which correlates with the background environment in which the sign is intended to be used but which simultaneously does not produce any glare or low visibility conditions.

When a safety sign or signal is capable of emitting both a steady and an intermittent output, the intermittent output should be used to convey the presence of a greater danger or greater requirement for remedial action than would be indicated if the output was continuous.

The Guidance on the Regulations (L64)[41] gives additional information.

A.3.5 Provision and Use of Work Equipment Regulations 1998[10]

Regulation 21 refers to lighting. Any place where a person uses work equipment should have lighting which is suitable and sufficient. If the ambient lighting installation in the workplace provides a suitable and sufficient environment for carrying out the visual tasks involved then special lighting need not be installed. If, however, the visual tasks involve the perception of fine detailed work, then additional lighting will be necessary in order to comply with the regulations.

When considering machinery, parts of which may be inadequately lit by normal lighting (by virtue of the peculiarities of the construction of the machine or any guards provided in the interests of safety) then additional local lighting may be required in order to give a sufficiently clear and unambiguous view of a dangerous process. Local lighting may also be required in order to reduce visual fatigue.

In work areas used for maintenance or repair, which are not covered by any general lighting provided, additional lighting should be provided. Such additional lighting could be temporary, for example the use of portable hand lamps, or it could be more permanent, e.g. by using fixed lighting inside enclosures.

The lighting requirements will be dictated by, inter alia, the purpose for which the work equipment is to be used and/or the nature of the work to be carried out. Due consideration should be given to the provision and installation of permanent lighting in areas where access is likely to be required on a routine basis.

This regulation complements the requirements given in the Workplace (Health, Safety and Welfare) Regulations 1992,[39] see A.3.2, in respect of suitable and sufficient workplace lighting. The Guidance on the Regulations (L22)[11] gives additional information.

A.3.6 Electricity at Work Regulations 1989[42]

Regulation 15 refers to working space, access and lighting. This regulation is designed to ensure, inter alia, that sufficient space, access and adequate illumination are provided whilst individuals are working on, at, or in the vicinity of, electrical equipment so that they may work in safety.

Daylight is preferred to artificial lighting, although where resort has to be made to artificial lighting, such an installation should be permanent and properly designed.

The memorandum on the Guidance to the Regulations (HS(R)25)[43] gives further information.

A.3.7 Electromagnetic Compatibility Regulations 1992[44]

Apparatus whose operation must not be affected by the working of other equipment includes lights and fluorescent lamps. Phenomena and effects which may be considered as causing an electromagnetic disturbance include:

- electric fields;
- magnetic fields;
- electromagnetic fields;
- oscillatory transients; and
- harmonics.

A.3.8 Docks Regulations 1988[32]

Regulation 6 refers to lighting. Each part of docks premises which is being used for normal loading, unloading and other associated work activities shall be lit suitably and adequately. Every obstruction and hazard on docks premises, which could present a potential problem when normal work activities are being carried out, shall be made conspicuous by one, or a combination, of several means including colouring, marking and lighting. The Approved Code of Practice with Regulations and Guidance[33] gives additional information.

A.3.9 Electrical Equipment (Safety) Regulations 1994[45]

Section 5 of the Regulations refers to the requirement for electrical equipment to be safe. Equipment shall be designed and constructed so that when connected to the mains electrical supply network system suitable protection against electric shock shall be provided by means of a combination of electrical insulation and the protective earthing conductor contained within the electricity supply system, or which achieves the same level of protection by other means.

Individuals and domestic animals must be provided with adequate protection against danger, injury, etc., which might be caused by either direct or indirect contact with electrical equipment.

A.3.10 Supply of Machinery (Safety) Regulations 1992[46]

The manufacturer of machinery must supply integral lighting which is appropriate for the operations likely to be undertaken when using the machinery, where its absence is likely to pose a risk even when working in ambient lighting which is considered to be of normal intensity. It is also necessary for the manufacturer to ensure that there is no area or shadow which is deemed likely to cause a nuisance, that there is no dazzle which would cause irritation and further that there are no dangerous stroboscopic effects likely to be created due to the lighting equipment supplied by the manufacturer.

Any internal assemblies of the machinery which will require routine inspection and maintenance must be provided with appropriate lighting. HSE Miscellaneous Information Sheet 9[47] gives further information.

A.3.11 Control of Substances Hazardous to Health Regulations 1994[17]

Section 7 refers to the prevention and control of exposure to substances hazardous to health and Section 10 refers to monitoring exposure at the workplace. Section 10 refers to the duty on employers to ensure that those in their employ are not subjected to exposure from substances hazardous to health, including solids, liquids and gases which are toxic, corrosive, harmful or irritating.

A.3.12 Special Waste Regulations 1996 (as amended)[19]

The waste produced either from accidental breakage of lamps or from the controlled breakage and crushing after lamps have expired must be disposed of in accordance with the conditions specified.

Factors which determine whether lamp waste is considered as special waste depends upon, inter alia, whether the lamps contain mercury or sodium and the quantity of the waste involved. Detailed information in respect of compliance with the Special Waste Regulations 1996[19] (as amended) can be obtained from the Environment Agency.[20]

A.4 Compliance with Health and Safety Law

In the event of a contravention of one or more of the laws of health and safety an inspector representing the Health and Safety Executive can take one of several forms of action:

- An *improvement notice* can be served if there is a contravention of any legislation, with the aim of carrying out the necessary corrective action within a specified time. This notice may be served on any individual who, in the opinion of the inspector, is contravening (or has contravened) any statutory legislation. Improvement notices may be served upon both employers and employees. If an individual is served with an improvement notice, that individual has the right to appeal against the notice, or any of the terms specified therein, to an industrial tribunal.
- A *prohibition notice* can be issued if there is a risk of serious personal injury. This is designed to stop the activity which gives rise to the risk until necessary corrective action, as specified in the notice, has been taken. The notice can be issued irrespective of whether there has been contravention of health and safety legislation and the date of effect of such notice can be immediate or at a later date. The notice can be served either on the individual carrying out the activity or alternatively on an individual in control of the activity at the time the notice was served.
- A *prosecution* can be initiated against a person who contravenes statutory legislation, in addition to the serving of a notice. Certain types of offence may be prosecuted only summarily, which involves a magistrate's court in England and Wales or in Scotland a sheriff's court. Many offences may be prosecuted either summarily, or alternatively on indictment in the crown court in England and Wales or in Scotland in the sheriff's court in solemn procedure. In many cases, where the offences alleged to have been committed can be tried in either of the methods described, they are prosecuted summarily if both the court and the defendant are agreeable. Imprisonment is an option open to the courts for those found guilty of certain offences against health and safety legislation.

It is normal for enforcing authorities to use discretion in deciding whether to initiate prosecution proceedings. Often other measures designed to achieve enforcement will initially be explored but, where the gravity of the situation so dictates, prosecution may be instigated without prior warning. In England and Wales the decision to proceed with a case in court normally rests with the enforcing authorities. In Scotland such decisions usually come under the jurisdiction of the Procurator Fiscal.

In addition to any other penalties which the court may impose, they may also make an order requiring the cause of the offence to be corrected. If an individual on whom either a prohibition notice or an improvement notice has been served does not comply with the terms specified in that notice, then they are liable to prosecution. Failure to comply with the terms specified in a prohibition notice could result in a custodial sentence being imposed.

- An inspector has the authority to seize, render harmless or destroy any material that is considered to be the likely cause of forthcoming danger or potential injury to individuals.

References

1. *Association of Optometrists (AOP) Handbook*, Association of Optometrists, London.
2. Precise Color Communication 9242-4830-92, Minolta (UK) Limited, Milton Keynes.
3. *Encyclopaedia of Occupational Health and Safety*, 4th edition 1998, Vol. II, ISBN 92 2 109815 X, International Labour Office (ILO), Geneva.
4. Smith, N.A. (PhD thesis) The steady-state and post-ignition transient luminous behaviour of the tubular fluorescent lamp operating throughout the dimmed range using high frequency dimming, University of Sheffield, 1996.
5. Smith N.A., *Lighting for Occupational Hygienists*, ISBN 0 948237 06 6, H & H Scientific Consultants, 1991.
6. International Electrotechnical Commission (IEC), IEC 825-1.
7. BS EN 60825-1 Safety of Laser Products, Part 1 Equipment Classification, requirements and user's guide.
8. Personal Protective Equipment at Work Regulations 1992, Statutory Instrument SI 2966.
9. Personal Protective Equipment at Work: Guidance on Regulations (L25), ISBN 0 7176 0415 2.
10. Provision and Use of Work Equipment Regulations 1998, Statutory Instrument SI 2306.
11. Guidance on Provision and Use of Work Equipment Regulations (L22), ISBN 0 7176 16266.
12. BS EN 207: 1994: Specification for filters and equipment used for personal eye protection against laser radiation.
13. BS EN 208: 1994: Specification for personal eye protection used for adjustment work on lasers and laser systems.
14. The radiation safety of lasers used for display purposes, Health and Safety Executive, HS (G) 95, ISBN 0717606910.
15. BS EN 60825-2: 1995: Safety of laser products. Safety of optical fibre communication systems.
16. International Electrotechnical Commission (IEC), International Lamp Coding System (ILCOS), 1993, IEC Document No. 123-93.
17. Control of Substance Hazardous to Health Regulations 1994, Statutory Instrument SI 3246.
18. Control of substances hazardous to health and control of carcinogenic substances: Control of Substances Hazardous to Health Regulations (L5) 1994: Approved Codes of Practice, ISBN 0 7176 0819 0.
19. Special Waste Regulations 1996 (as amended) Statutory Instrument SI 972.
20. Environment Agency.
21. BS 4533: Luminaires.
22. BS 5345: 1989: Code of practice for selection, installation, and maintenance of electrical apparatus for use in potentially explosive atmospheres (other than mining applications or explosive processing and manufacture).

23. BS 6467: 1988 Part 2: Electrical apparatus with protection by enclosure for use in the presence of combustible dusts.
24. BS EN 60598 Luminaires.
25. Building Regulations 1991, Approved Document B, ISBN 0 11 752313 5.
26. BS 5266: 1988: Emergency Lighting, Part 1 Code of practice for the emergency lighting of premises other than cinemas and certain other specified premises used for entertainment.
27. CIBSE LG4: Sport 1990, ISBN 0 900953 45 4, Chartered Institution of Building Services Engineers (CIBSE), London.
28. BS 5489, Road Lighting.
29. Code for Interior Lighting 1994, ISBN 0 900953 64 0, Chartered Institution of Building Services Engineers (CIBSE), London.
30. BS 950, Part 1: 1967, Specification for artificial daylight for the assessment of colour.
31. *Lighting in Printing Works*, ISBN 0 85168 131 X, British Printing Industries Federation, 1980.
32. Docks Regulations 1988, Statutory Instrument 1655.
33. COP25 Safety in docks: Docks Regulations 1988, Approved Code of Practice with Regulations and Guidance, ISBN 0 11 885456 9.
34. Health and Safety (Display Screen Equipment) Regulations 1992, Statutory Instrument SI 2792.
35. Health and Safety at Work etc. Act 1974.
36. Display screen equipment work: Health and Safety (Display Screen Equipment) Regulations 1992: Guidance on Regulations (L26), ISBN 0 7176 0410 1.
37. Management of Health and Safety at Work Regulations 1992, Statutory Instrument SI 2051.
38. Management of Health and Safety at Work: Approved Code of Practice (L21), ISBN 0 7176 0412 8.
39. Workplace (Health, Safety and Welfare) Regulations 1992, Statutory Instrument 3004.
40. Health and Safety (Safety Signs and Signals) Regulations 1996, Statutory Instrument 341.
41. Safety signs and signals: the Health and Safety (Safety Signs and Signals) Regulations 1996, Guidance on Regulations (L64), ISBN 0 7176 0870 0.
42. Electricity at Work Regulations 1989, Statutory Instrument 635.
43. The memorandum of guidance on the Electricity at Work Regulations 1989 (HS(R)25), ISBN 0 11 883963 2.
44. Electromagnetic Compatibility Regulations 1992, Statutory Instrument SI 2372.
45. Electrical Equipment (Safety) Regulations 1994, Statutory Instrument SI 3260.
46. Supply of Machinery (Safety) Regulations 1992, Statutory Instrument SI 3073.
47. HSE Miscellaneous Information Sheet 9, Supply of Machinery (Safety) Regulations 1992, DTI Booklet URN 95/650 Machinery: Guidance Notes on UK Regulations.

Suggested further reading

1. CIBSE LG1: Lighting Guide, The Industrial Environment, ISBN 0 900953 38 1.
2. CIBSE LG2: Lighting Guide, Hospitals and Health Care Buildings, ISBN 0 900053 37 3.
3. CIBSE LG3: Lighting Guide, Areas for Visual Display Terminals, ISBN 0 900953 41 1.
4. CIBSE Lighting for Offices 1993, ISBN 0 900953 63 2.
5. CIBSE TM10: Technical Memorandum, Calculation of Glare Indices, 1985.
6. Lighting Industry Federation (LIF), LIF Technical Statements 1 to 17.
7. *Occupational Hygiene*, edited by K. Gardner and J.M. Harrington (chapter on Lamps and Lighting by N.A. Smith), ISBN 0 632 03734 2, Blackwell Scientific Publications, Oxford, 1995.
8. Workplace Lighting – Special Report – Issue 25 1996, Report by N.A. Smith, ISSN 0967 8344, Croner Publications Ltd.
9. *The Workplace*, Volume 1, (section by N.A. Smith), ISBN 82 91833 00 1, International Occupational Safety and Health Information Centre (Geneva) and Scandinavian Science Publisher (Oslo), 1997.
10. *Lighting for Occupational Optometry*, Smith, N.A., ISBN 0 948237 35 X, H & H Scientific Consultants, 1999.
11. CIBSE TM12: Technical Memorandum, Emergency Lighting, 1986.
12. CIBSE Application Guide, Hostile and Hazardous Environments, ISBN 0 900953 26 8.
13. BS 5501, Electrical apparatus for potentially explosive atmospheres.
14. IEC 79 Electrical apparatus for explosive gas atmospheres.
15. EN 55 015 Radio interference limits (fluorescent lamps).
16. Health and Safety Executive, HSE 253/3 Disposal of Discharge Lamps.
17. Environmental Protection Act, 1990.
18. Control of Pollution (Amendment) Act, 1989.
19. Trade Effluent Regulations 1989, SI 1156.
20. The Health and Safety Executive, Working with Employers HSE26.
21. Guidance Notes for the Reduction of Light Pollution, The Institution of Lighting Engineers, 1997.

Index